120 Years of Nitrate Therapy –
Prepared for the Next Millennium

With compliments

SCHWARZ
PHARMA

120 Years of Nitrate Therapy – Prepared for the Next Millennium

Edited by

Ch.J.F. Holubarsch · T.F. Lüscher

Walter de Gruyter
Berlin · New York 2000

Editors

Thomas F. Lüscher, MD
Universitätsspital Zürich, Division of Cardiology
Rämistrasse 100, 8091 Zürich, Switzerland

Christian J.F. Holubarsch, MD
Medizinische Universitätsklinik Freiburg, Department of Cardiology and Angiology
Hugstetter Strasse 55, 79106 Freiburg, Germany

Die Deutsche Bibliothek – Cataloging in Publication Data

120 years of nitrate therapy – prepared for the next millennium / ed. by Ch.J.F. Holubarsch ; T.F. Lüscher. – Berlin ; New York : de Gruyter, 2000
ISBN 3-11-016848-0

Medical science is constantly developing. Research and clinical experience expand our knowledge, especially with regard to treatment and medication.

For dosages and applications mentioned in this work, the reader may rely on the authors, editors and publishers having taken great pains to ensure that these indications reflect the standard of knowledge at the time this work was completed.

Nevertheless, all users are requested to check the package leaflet of the medication, in order to determine for themselves whether the recommendations given for the dosages or the likely contraindications differ from those given in this book. This is especially true for medication which is seldom used or has recently been put on the market and for medication whose application has been restricted by the German Ministry of Health.

The quotation of registered names, trade names, trade marks, etc. in this book does not imply, even in the absence of a specific statement that such names are exempt from laws and regulations protecting trade marks, etc. and therefore free for general use.

Copy-editing, reproductions, typesetting: K. Handwerker, Wissenschafts-Lektorat & DTP Service, Berlin.
Cover design: Rudolf Hübler, Berlin.
Printing: Gerike GmbH, Berlin. – Binding: Lüderitz & Bauer, Berlin.
Printed in Germany

Contributors

Tim Baxter, MD
Schwarz Pharma Ltd., Schwarz House, East Street, Chesham, Buckinghamshire HPG5 1DG, Great Britain

Ton J.M. Cleophas, MD
Albert Schweitzer Hospital, P.O.Box 306, 3300 AH Dordrecht, The Netherlands

Gad Cotter, MD
Assafh-Harofeh Medical Center, The Cardiology Institute, 70300 Zerifin, Israel

Stefano Ghio, MD
University of Pavia, Policlinico "San Matteo", Department of Cardiology, 27100 Pavia, Italy

Christian J.F. Holubarsch, MD
Medizinische Universitätsklinik Freiburg, Department of Cardiology and Angiology, Hugstetter Strasse 55, 79106 Freiburg, Germany

Louis J. Ignarro, PhD
UCLA School of Medicine, Department of Molecular and Medical Pharmacology, Center for Health Sciences, Los Angeles, CA, USA

Rutger M.G. Jansen, MSc
Martini Hospital, Department of Cardiology, P.O.Box 30033, 9700 RM Groningen, The Netherlands

Mazhar Khan, MD
Royal Victoria Hospital, Regional Medical Cardiology Center, Grosvenor Road, Belfast BT12 6BA, Northern Ireland

Thomas F. Lüscher, MD
Universitätsspital Zürich, Division of Cardiology, Rämistrasse 100, 8091 Zürich, Switzerland

Menco G. Niemeyer, MD, PhD
Martini Hospital, Department of Cardiology, P.O.Box 30033, 9700 RM Groningen, The Netherlands

Georg Noll, MD
Universitätsspital Zürich, Division of Cardiology, Rämistrasse 100, 8091 Zürich, Switzerland

Axel Rehe, PhD
Schwarz-Pharma AG, Alfred-Nobel-Strasse 10, 40789 Monheim, Germany

Frank Ruschitzka, MD
Universitätsspital Zürich, Division of Cardiology, Rämistrasse 100, 8091 Zürich, Switzerland

Claudia Raineri, MD
University of Pavia, Policlinico "San Matteo", Department of Cardiology, 27100 Pavia, Italy

Ph. Gabriel Steg, MD
Hôpital Bichat, Service de Cardiologie A, 46, Rue Henri Huchard, 75877 Paris Cedex 18, France

Aeilko H. Zwinderman, PhD
Faculteit Geneeskunde, Vakgroep MESIB, Sectie Medische Statistiek, Postbus 9604, 2300 RC Leiden, The Netherlands

Contents

Introduction
Ch.J.F. Holubarsch, T.F. Lüscher 1

Role of nitric oxide as a signalling molecule in the cardiovascular system
L.J. Ignarro 7

From an explosive to the endogenous nitrate: historial, physiological and clinical aspects
T.F. Lüscher, F. Ruschitzka, G. Noll 19

The role of nitrates in the therapy of congestive heart failure
Ch.J.F. Holubarsch 33

Long-term oral treatment with nitrates – what is important?
M. Khan 43

Alternatives in long-term nitrate therapy
S. Ghio, C. Raineri 55

The Yin and Yang of nitric oxide, or: the importance of vascular control in acute ischaemia and decrease in cardiac output
G. Cotter 61

Economic impact of Elantan LA compared to Tenormin and Tildiem LA in the treatment of stable angina in the UK
T. Baxter, A. Rehe 75

Quality of life in angina therapy: focus on the beneficial effects of nitrates
M.G. Niemeyer, R.M.G. Jansen, T.J.M. Cleophas, A.H. Zwinderman 89

New trends in interventional cardiology
P.G. Steg 99

Introduction

Ch.J.F. Holubarsch, T.F. Lüscher

"It is a non-smelling, clear to browny, fluid substance produced by nitrilation of glycerin, which is stable at room temperature but may explode on occasion of a push or hit." This is the simple description of nitroglycerin in a textbook of chemistry (molecular structure see Figure 1).

Nitroglycerin has two very different and distinguished histories, one as an explosive and one as a remedy.

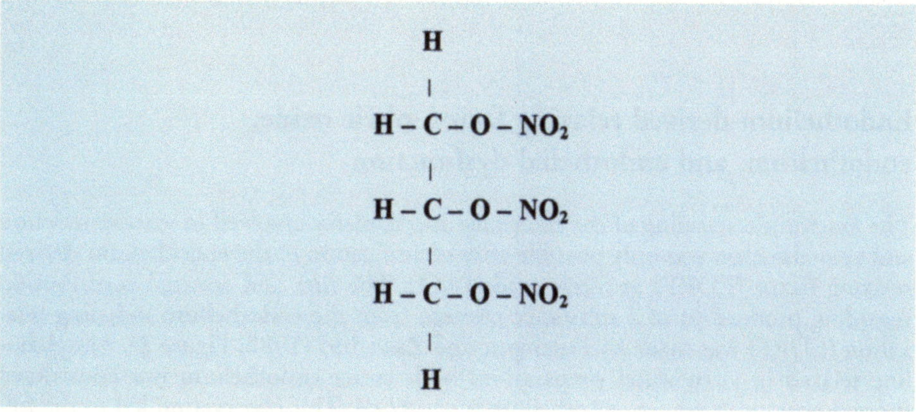

Figure 1: Molecular structure of nitroglycerin.

Nitroglycerin as an explosive

Alfred Nobel (1833–1896), a Swedish chemist and industrial professional, was the owner of the patent for production of nitroglycerin. He built factories for nitroglycerin production in Sweden and Germany starting 1864. Interestingly, his experience with nitroglycerin was initially traumatizing, since one of his factories exploded in 1884. In those days, the first symptoms of nitrate tolerance were reported by workers in nitroglycerin factories. They experienced typical headaches, particularly on Mondays when they started to work with the substance again after a free weekend.

Alfred Nobel donated his property to the Nobel Foundation from which – each year – the Nobel Prizes have been paid since 1901.

Nitroglycerin as a remedy

The first to describe the beneficial effects of nitroglycerin in humans was Thomas Lauder Brunton in 1867. At that time, he used nitrate of amyl to treat symptoms of angina. Since then, nitroglycerin and other nitrates have been increasingly used in patients with coronary heart disease and heart failure up to the present day.

During the last 20 years, our knowledge of the function and mechanism of nitroglycerin has grown enormously, since nitric oxide was shown to be the active molecule that provides vasodilation and is produced by the endothelium of the organism itself (endothelium-derived relaxing factor (EDRF) = nitric oxide (NO)).

Endothelium-derived relaxing factor, nitric oxide, endothelium, and endothelial dysfunction

The exact understanding of the molecular mechanisms involved in vasoconstriction and vasorelaxation was only possible after identification of the endothelium-derived relaxing factor (EDRF) as nitric oxide (NO): The first and seminal contribution regarding production of a substance released from the endothelium inducing relaxation (EDRF) was made by Furchgott and Zawadzki (1980, Figure 2). Acetylcholine relaxed in vitro vessel preparations with intact endothelium but constricted those preparations whose endothelium was rubbed. This observation led to the following hypothesis: acetylcholine has two effects in the intact vessel: (1) Acetylcholine has a direct vasoconstriction effect on vascular smooth muscle cells (rubbed endothelium in Figure 2) (2). Acetylcholine stimulates the endothelium to release a relaxing factor, (EDRF), which diffuses to the smooth muscle cells and induces relaxation (intact endothelium in Figure 2; left side of Figure 3). In the intact vessel, the net effect of acetylcholine (relaxation minus constriction) is vasorelaxation (Figure 2).

It was the work of L.J. Ignarro and coworkers as well as of Robert Furchgott to identify EDRF as nitric oxide (1987). Only after these basic scientific discoveries were clinical trials initiated to study the influence of acetylcholine in addition to nitroglycerin on the function of coronary arteries in humans. In a number of highly specialized catheter laboratories it was demonstrated that acetylcholine relaxes coronary vessels in the same way as nitroglycerin in healthy subjects. However, in those patients suffering from coronary artery disease or in those with risk factors like arterial hypertension, hypercholesterinaemia and smoking, acetylcholine produces paradoxical effects in the sense of vasoconstriction. It also became evident that both

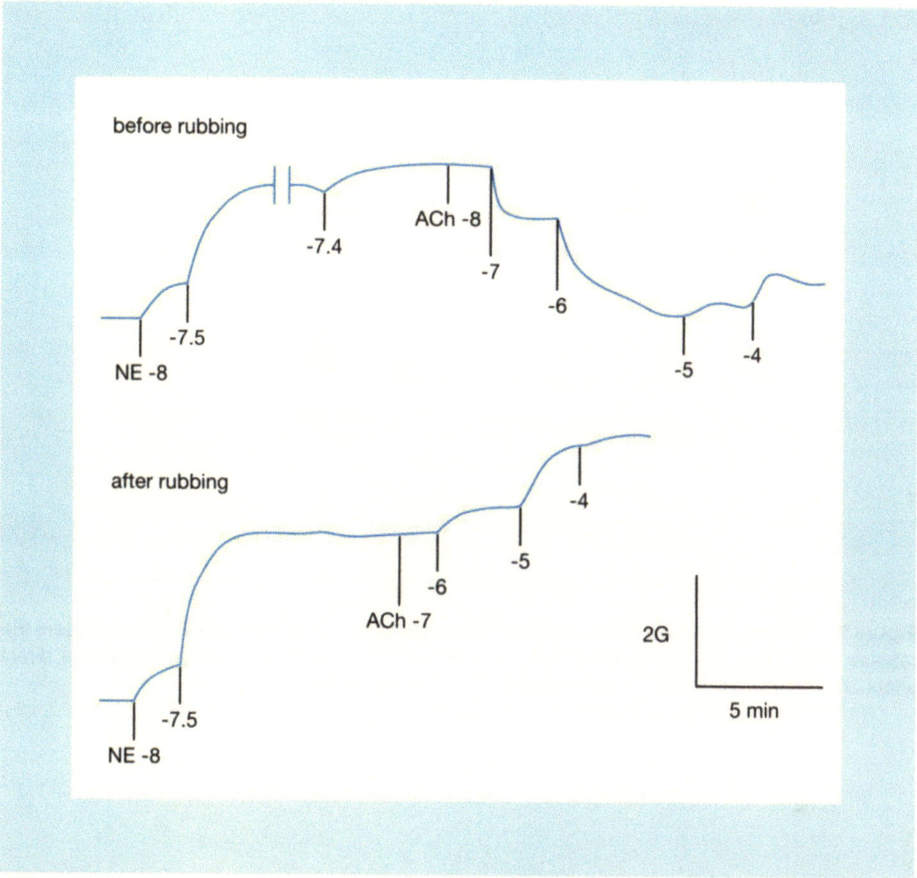

Figure 2: Effect of acetylcholine (ACh) on vessel preparations with and without intact endothelium. Precontraction was performed by addition of norepinephrine (NE) (according to Furchgott and Zawatzki, 1980).

large vessels and the microcirculation are involved in endothelial dysfunction. These observations indicate that the effect of acetylcholine on the endothelium to release EDRF = NO is blunted in the same way. Newer studies, however, give evidence for the hypothesis that – at least under certain conditions – nitric oxide is produced and released quite properly but nitric oxide may be inactivated by free oxygen radicals.

Free oxygen radicals are typically eliminated by superoxide-dismutase (SOD) and catalase. However, if the concentrations of oxygen radicals are too high, they react with nitric oxide to produce peroxynitrate. This substance is not only a weak vasodilator compared to nitric oxide but is also cytotoxic (as it produces nitrotyroxine residues in vital enzymes) and thereby contributes to the progression of arteriosclerosis.

Important therapeutic goals to treat endothelial dysfunction, and thereby atherosclerosis, must therefore be either the prevention of production of oxygen radicals or the inactivation of oxygen radicals.

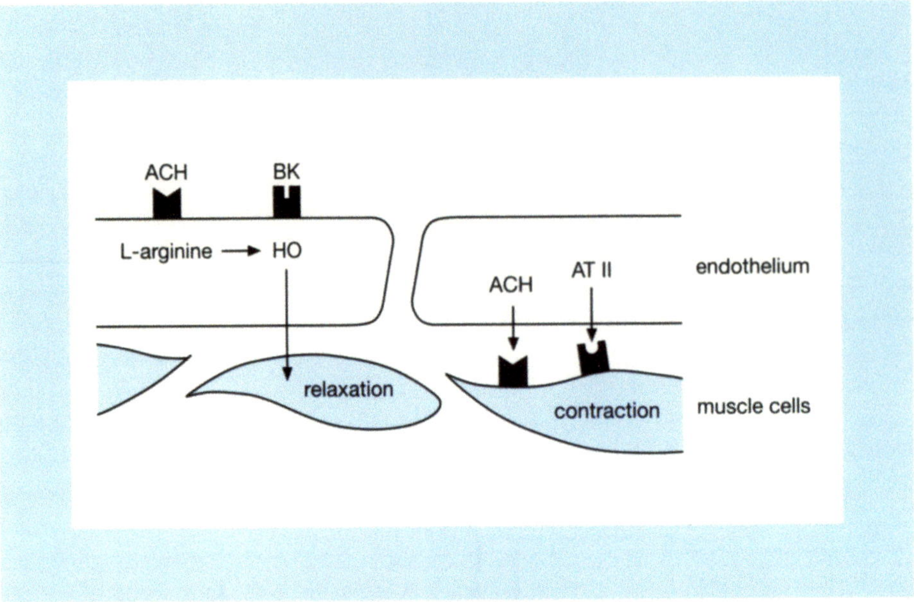

Figure 3: Explanation of the different effects of acetylcholine in Figure 2: Acetylcholine triggers the release of NO from the endothelium (left, vasorelaxation). In addition, acetylcholine has a direct effect on smooth muscle cells in the sense of vasoconstriction (right).

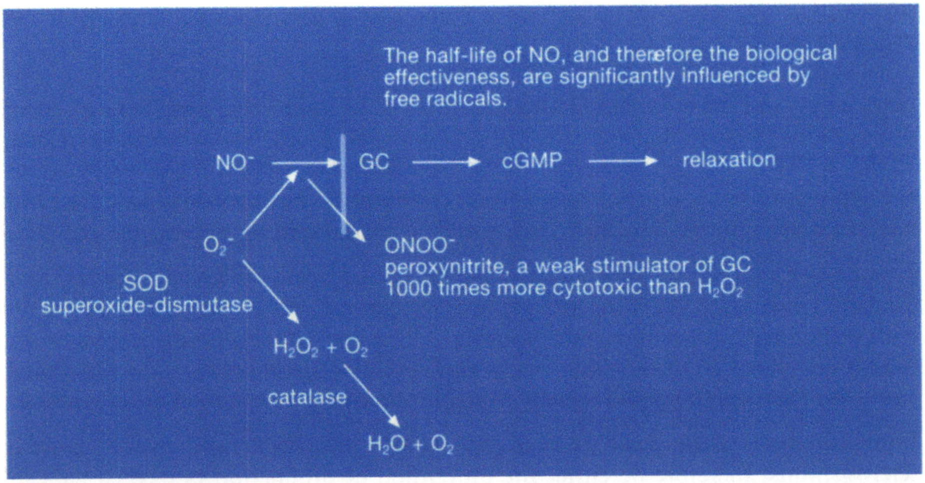

Figure 4: Inactivation of nitric oxide by free oxygen radicals.

Whereas the former goal may be reached by application of ACE-inhibitors or angiotensin-receptor antagonists – angiotensin stimulates membrane-bound oxidases to produce oxygen radicals – oxygen radicals may be scavangered at least in part by concomitant application of antioxidants like vitamin C.

Nitroglycerin tolerance

Continuous application of nitroglycerin leads to the development of tolerance within less than 24 hours. Like endothelial dysfunction, the most plausible explanation of this phenomenon is the production of oxygen radicals. To prevent nitrate tolerance, either intermittent application of nitroglycerin is indicated, or addition of an ACE-inhibitor or angiotensin receptor antagonist may attenuate the tolerance problem.

The symposium

Since nitroglycerin was first used as a remedy to treat angina 120 years ago, Schwarz Pharma sponsored an international symposium entitled "120 years of nitrate therapy – prepared for the next millennium", which was held in Berlin September 17 to 19, 1999. The organizers were proud and the two hundred of participants were pleased to listen to – in addition to other famous scientists – the Nobel Prize laureate from 1998, Dr. Louis J. Ignarro.

Role of nitric oxide as a signalling molecule in the cardiovascular system

L.J. Ignarro

Nitric oxide as a smooth muscle relaxant

The ability of nitric oxide (NO) to relax vascular smooth muscle was discovered about 20 years ago (1). The original observation was made by bubbling a gaseous mixture of nitric oxide in nitrogen into a tissue bath containing isolated precontracted strips of bovine coronary artery. The marked but transient arterial relaxation was blocked by haemoglobin, methaemoglobin and myoglobin and by the oxidant methylene blue. Nitric oxide also activated soluble guanylate cyclase from bovine coronary artery.

The pharmacological profile of nitroglycerin, nitroprusside, other organic nitrate esters and some organic nitrite esters led to the supposition that these agents all caused vasodilatation by acting as NO donor agents on contact with tissues in aqueous solution.

The extraordinarily high potency of nitroglycerin as a smooth muscle relaxant led scientists to postulate that tissue receptors for nitroglycerin exist because there must be an endogenous nitroglycerin or similar NO donor or nitric oxide itself in mammalian tissues. (It was not appreciated at that time that the vascular endothelium generated NO and delivered it to the underlying smooth muscle cells.)

Shortly after the conduction of these experiments on nitric oxide, Furchgott and Zawadzki discovered that acetylcholine-elicited relaxation of isolated arteral preparations is dependent on the presence of an intact endothelial layer (2). The unstable relaxing substance was also thought to be responsible for the arterial relaxation seen in response to bradykinin, histamine, ADP and ATP and this substance was termed endothelium-derived relaxation factor (EDRF).

In the 1980s EDRF was shown to relax smooth muscle by elevating cyclic guanosine monophosphate levels and to stimulate soluble guanylate cyclase, suggesting that EDRF might be nitric oxide (3).

In searching for the mechanism of vasodilator action, it was proposed that agents such as nitroglycerin either spontaneously released nitric oxide in aqueous solution or reated with tissue thiols to generate chemically unstable intermediates, S-nitrosothiols, that decomposed with the liberation of nitric oxide (4). For example, thiols such as cysteine and glutathione markedly enhanced the activation of guanylate cyclase by nitrite and cysteine is required for enzyme activation by nitroglycerin. Vasorelaxation induced by S-nitrosothiols was accompanied by tissue cyclic GMP formation and the profile of haemodynamic effects of the S-nitrosothiols was virtually identical to that of nitroglycerin. Two S-nitrosothiols, S-nitroso-N-acetylpenicillamine (SNAP) and S-nitrosoglutathione (GSNO) are widely used as NO donor agents.

This work led to a better understanding of the mechanism by which tolerance to the vasodilator action of glyceryl trinitrate can develop (with little or no cross-tolerance to nitroprusside or nitrite). Tissue thiols are required for chemical reaction with nitroglycerin to liberate nitric oxide from the intermediate S-nitrosothiols and activate soluble guanylate cyclase. Repeated administration of large doses of nitroglycerin leads to the depletion or oxidation of tissue thiols and the gradual diminution of action of nitroglycerin. The responsiveness of tolerant arterial smooth muscle to GTN can be restored by treatment of arterial tissue with sulfhydryl reducing agents (5).

EDRF or endothelium-derived nitric oxide is continuously generated from vascular endothelial cells in the absence of added endothelium-dependent vasodilators. Early clues for this basal release of nitric oxide came from the finding that vascular tissue cyclic GMP levels were significantly higher in endothelium-intact preparations than in endothelium-denuded preparations and from studies with haemoglobin and cyclic GMP phosphodiesterase inhibitors. Direct evidence for the basal generation of nitric oxide came from bioassay cascade experiments using perfused and/or superfused vascular preparations; here the continuous formation and release of nitric oxide could be directly observed (6).

The basal formation of nitric oxide varies between vessel types and across differing diameters in the same vessel segment eg greater basal nitric oxide formation is seen in bovine intrapulmonary artery and vein with smaller rather than larger diameters. The smaller vessel rings contain higher resting levels of cyclic GMP and it is more difficult to maintain a steady level of tone in these preparations.

The hypothesis was put forward that basal formation of arterial or venous nitric oxide may be important for the continuous modulation of vascular smooth muscle tone and also for platelet adhesion and aggregation (7). (Interference with basal production of nitric oxide using NO synthase inhibitors causes a prompt and sustained rise in systemic blood pressure.) (8) Endothelium-derived nitric oxide could counteract localised vasospasm and thrombus formation that could develop in response to factors released from vascular endothelium, activated platelets and other blood cells.

The shear stress generated by flowing blood against the endothelial cell surface triggers the generation of nitric oxide in endothelial cells. Shear forces trigger the opening of calcium channels on endothelial cells, leading to the calcium-dependent activation of NO synthase and increased local production of nitric oxide (9, 10). The increase in intracellular calcium concentration may be the result of tyrosine phosphorylation and activation of phospholipase C and protein phosphatases. Calcium-independent activation of endothelial NO synthase in response to shear stress can also occur. This may involve tyrosine phosphorylation of endothelial nitric oxide synthase or the action of another regulatory protein.

Nitric Oxide (NO) = endothelium-derived relaxation factor (EDRF)

Little more than a decade ago, experimental evidence from several laboratories showed that EDRF had similar pharmacological, biochemical and chemical properties to nitric oxide. For example, EDRF from artery and vein could activate purified soluble guanylate cyclase, and this is only activated by a limited number of substances (11). This activation was heme-dependent for both EDRF and NO. Also it was discovered in 1986 that EDRF inhibits platelet aggregation, like NO (12).

Proof that EDRF and NO have identical biological and chemical properties came in 1987 (13, 14) (Figure 1). Studies using pulmonary artery and vein showed that EDRF and NO both require guanylate cyclase-bound heme for enzyme activation, bind to a common site on haemoglobin and myoglobin, are inactivated by superoxide anion and protected by superoxide dismutase and react with haemoglobin to form the same adduct.

In another study the vascular effects of EDRF released from perfused bovine intrapulmonary artery and vein were compared with the effects of nitric oxide delivered by superfusion over endothelium-denuded arterial and venous strips arranged in a cascade. EDRF was indistinguishable from NO and was identified chemically as NO by two procedures: both substances reacted with haemoglobin to form nitrosylhaemoglobin and both similarly promoted the diazotization of sulfanilic acid and yielded the same reaction product after coupling with N-(1-naphthyl)-ethylenediamine.

The final proof was the demonstration, using simultaneous chemical assay and bioassay, that nitric oxide was released from endothelial cells by bradykinin in amounts that accounted for the actions of EDRF (15).

NO and platelet function

While conducting experiments into the vasorelaxant properties of nitric oxide, it was discovered that NO also inhibits platelet aggregation. For instance, glyceryl trinitrate (which elevates smooth muscle cyclic GMP levels and activates guanylate cyclase in coronary arteries) was observed to inhibit platelet aggregation by a mechanism unrelated to cyclic AMP formation (16).

Nitric oxide and a series of S-nitrosothiols, which are used as nitric oxide donor agents, were found to inhibit platelet aggregation, markedly activate soluble platelet guanylate cyclase (70–280 fold) and markedly elevate platelet cyclic GMP levels (100–200 fold) (17). Inhibition of platelet aggregation occurred regardless of the agent used to promote aggregation, whether ADP, collagen, thrombin or thromboxane analogue.

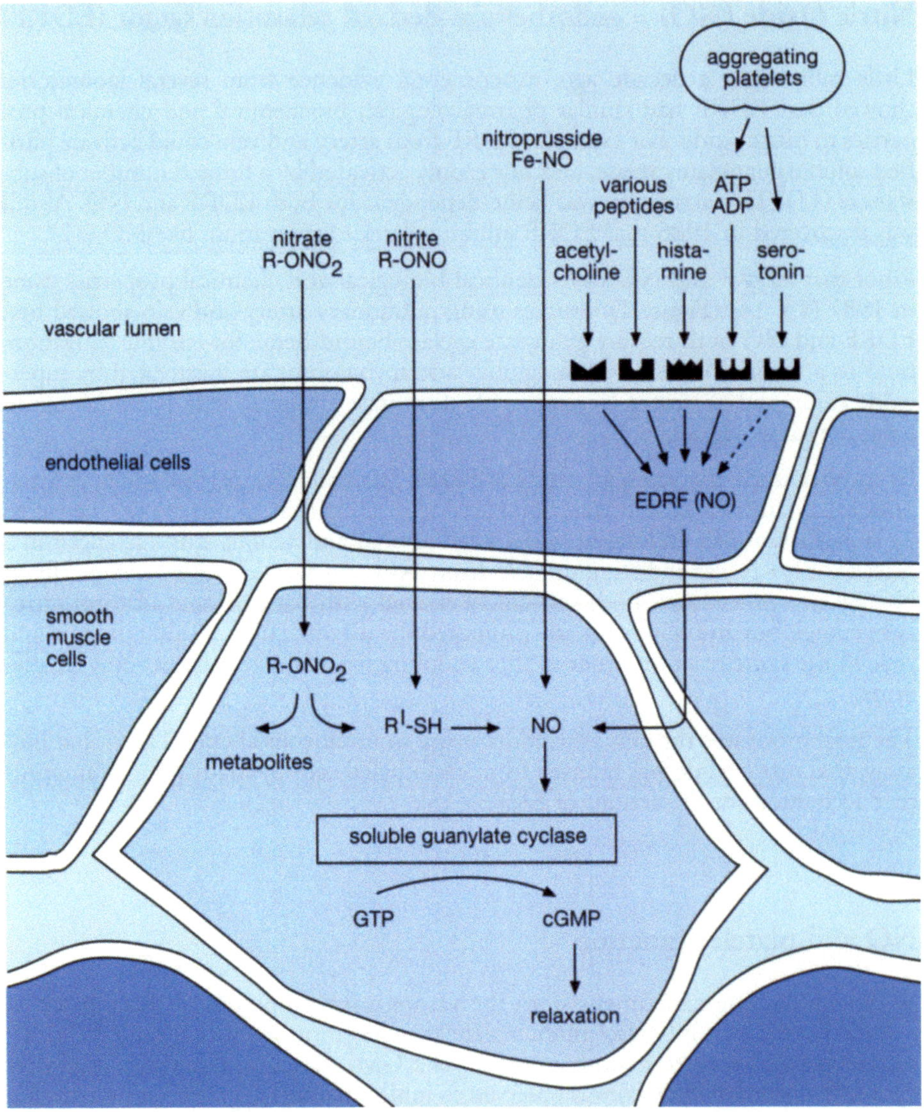

Figure 1: Schematic illustration of proposed mechanisms by which nitric oxide (NO) and nitrovasodilatators relax vascular smooth muscle.

In addition to preventing the onset of platelet aggregation, nitric oxide caused platelet disaggregation that was preceded by cyclic GMP accumulation. The anti-aggregatory effects of nitric oxide are synergistic with those of another endothelium-derived mediator, prostacyclin.

Since platelet aggregation may play a part in a variety of thromboembolic disorders – myocardial infarction, cerebrovascular disease and atherosclerosis – prevention and/or reversal of platelet aggregation by drugs by mechanisms involving cyclic GMP formation might provide some therapeutic opportunities.

Mechanism of action of nitric oxide

Nitric oxide was first reported to activate soluble guanylate cyclase in the mid-1970s (18). The effects of heme on guanylate cyclase activation were studied using heme-deficient and heme-containing forms of the enzyme. These experiments showed that the presence of heme is obligatory for the activation of guanylate cyclase by NO and by agents that release or generate NO such as nitroprusside, inorganic nitrite and hydroxylamine. Using purified guanylate cyclase, heme was found to be bound to the enzyme as a prosthetic group (19).

The hypothesis was put forward that NO binds to the heme iron to form the nitrosyl-heme adduct of guanylate cyclase. The heme was thought to bind as a 5-coordinate complex, with the fifth bond being the axial ligand between heme iron and histidine in the enzyme protein. On interaction of the heme iron with nitric oxide, the axial ligand undergoes cleavage, leading to projection of heme iron away from the enzyme protein and out of plane of the porphyrin ring configuration. This configurational change would also modify the nearby active catalytic site to increase access to the surface where magnesium and GTP bind.

This hypothesis would explain why activation of guanylate cyclase by nitric oxide causes a 200–400-fold increase in Vmax and a three-fold decrease in the Km for MgGTP (20). This property makes guanylate cyclase activatable only by nitric oxide; the cyclic GMP system has great selectivity since only NO can stimulate significant cyclic GMP production in tissues.

Soluble guanylate cyclase also contains bound copper. Its role may be to facilitate the release of nitric oxide from S-nitrosothiols, thereby facilitating the activation of guanylate cyclase (Figure 2).

Isoforms of nitric oxide synthase

Neuronal NO synthase (nNOS)

Ten years ago certain areas of the brain were found to contain high levels of cyclic GMP and cytosolic guanylate cyclase (21). Arginine is the soluble endogenous activator of guanylate cyclase (22). NMDA stimulates cyclic GMP levels in rat cerebellar cells, releasing an EDRF-like factor which is in fact nitric oxide. The arginine to NO pathway is described as the NO synthase enzymatic pathway and the relevant enzyme as neuronal nitric oxide synthase (nNOS). In the central nervous system NO has been implicated in long-term potentiation in the hippocampus, as

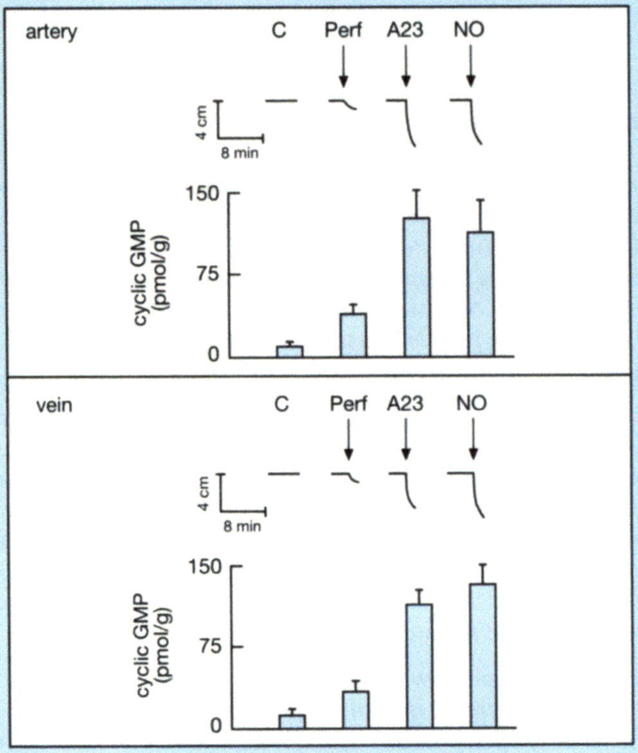

Figure 2: Similarity between EDRF release from perfused artery and vein and superfused NO on Cyclic GMP accumulation in vascular smooth muscle. Endothelium-denuded strips of pulmonary artery or vein, respectively, were precontracted with 10 µM phenylephrine or 0,01 µM U46619 delivered by superfusion. Two vascular strips were arranged in a cascade, and the time delay of superfusion between the strips was adjusted to 0,3–0,5 sec. Control © strips signify absence of arterial or venous perfusion or superfusion with NO, and srips were quick-frozen at the time of peak contractile response. Perf signifies perfusate from the intact segment of artery or vein, and strips were quick-frozen 2 min after vessel perfusates were allowed to superfuse the strips. A23187 (1 µM; A23) was perfused through the vessel segmants, whereas NO (1 µM) was superfused over the strips, and the strips were quick-frozen 2 min after onset of relaxation.

a mediator of short-term electrocortical activation and as a modulator of pain perception.

In the peripheral nervous system, NO appears to function as a neurotransmitter mediating vascular and non-vascular smooth muscle relaxation. These tissues are innervated by nonadrenergic-noncholinergic (NANC) neurons. NANC neurons exist in the outer adventitial layers of large blood vessels and in arteries within erectile tissue. In the corpus cavernosum of the penis, NO mediates penile erection by relaxing smooth muscle, allowing the vascular beds to become engorged with blood (23). NNOS-containing neurons are found in the myenteric plexus of the gastrointestinal tract, where they are involved in peristalsis (24).

Nearly all organs in the mammal are innervated by neurons that release nitric oxide as a neurotransmitter, which elicit both physiological and pathophysiological actions. There is increasing evidence that nNOS can undergo expressional regulation. Upregulation of nNOS mRNA represents a general response of neuronal cells to stress induced by a wide variety of physical, chemical and biological agents and may ultimately result in cellular injury and apoptosis (25).

Inducible NO synthase (iNOS)

Inducible nitric oxide synthase is not a normal constituent of quiescent cells but is expressed after cells are activated by exposure to products of infection, lipopolysaccharide or inflammatory cytokines (26). For example, activated macrophages synthesise nitric oxide from arginine and subsequently oxidise it to nitrite and nitrate. Formation of NO by iNOS has been demonstrated in many tissues, including vascular smooth muscle and endothelial cells, hepatocytes, fibroblasts and astrocytes.

The high-output production of NO catalysed by iNOS results in the initiation and/or exacerbation of symptoms of acute and chronic inflammation. Inhibitors of iNOS catalytic activity produce anti-inflammatory effects. Nitric oxide is an important host-defence mechanism, being involved in the killing of pathogens such as mycobacterium tuberculosis and malaria parasites and it also regulates lymphocyte function (27).

Endothelial NO synthase (eNOS)

A nitric oxide synthase is present as a normal constituent of healthy endothelial cells. It produces nitric oxide that acts as an intra- and inter-cellular messenger molecule in response to increases in intracellular calcium concentration.

The NOS isoforms catalyse the oxidation or oxygenation of arginine to nitric oxide plus citrulline (28). This is a cytochrome P450-like oxidation reaction and requires NADPH as the principal electron donor. Like cytochrome P450, NOS contains heme as a prosthetic group for oxygen binding and substrate formation. NOS is also a flavoprotein and requires both FAD and FMN for the full expression of catalytic activity. Calmodulin, a calcium binding protein, facilitates the transfer of electrons from enzyme-bound FAD to the heme iron and acts as an on/off switch for NO biosynthesis. Tetrahydrobiopterin is also required for full NOS catalytic activity; it acts by stabilising the NOS protein.

There is good homology of the C-terminal portions of the three isoforms – nNOS, iNOS and eNOS – and they all show close homology to NADPH cytochrome

P450 reductase (29). The C-terminal reduction domain contains binding sites for FMN, FAD and NADPH whereas the N-terminal oxygenase domain contains binding sites for heme, tetrahydrobiopterin and L-arginine. The oxygenase and reductase domains are linked by a chain of amino acids containing a calmodulin-binding site. The two domains can fold and function independently of one another.

NO as an ideal modulator

Nitric oxide has many characteristics which make it able to act as a local modulator of blood flow, haemostasis and platelet function. Since it originates in endothelial vascular cells it can be delivered to the underlying smooth muscle and to the endothelial surface where it can interact with adhering or nearby platelets. Its small size and lipophilic nature enable it to diffuse rapidly through cell membranes and its chemical instability gives a highly local activity. Thus nitric oxide has the unique capacity to engage in cell-to-cell communication and can mediate, for example, improved local blood flow to an injured tissue.

Physiological control mechanisms may act to diminish nitric oxide production. For example, NO itself can act as a negative feedback modulator of NO synthase catalytic activity through an interaction between NO and the heme iron in nitric oxide synthase.

Nitric oxide in disease and possible drug development

Decreased synthesis or action of nitric oxide is implicated in a wide variety of disease processes (30). For instance, decreased vasodilatation is seen in patients with hypertension, diabetes mellitus, atheroma and hypercholesterolaemia. Nitric oxide may be involved in coronary vasospasm in atherosclerotic arteries and cerebral vasospasm after subarachnoid haemorrhage. Increased NO production in the heart may contribute to cardiomyopathy and ventricular dysfunction. It is implicated in cerebral damage caused by stroke, Parkinson's disease and AIDS dementia.

In renal failure, accumulation of ADMA (an endogenous inhibitor of nitric oxide synthesis) could contribute to hypertension or immune dysfunction. At least one form of impotence may be attributable to a defect in the arginine-NO-cyclic GMP pathway so that diminished relaxation of corpus cavernosum smooth muscle occurs after stimulation of the NANC nerves.

These data encourage the development of new drugs, based on our understanding of the functions of nitric oxide, as vasodilators and antiplatelet agents in the treatment of hypertension, stroke, angina, heart failure, the vascular complications of diabetes mellitus, impotence and other vascular disorders. An excellent example of this potential is the development of sildenafil (Viagra) for the treatment of impotence.

Abnormalities of nitrergic neurons in the airways could increase bronchoconstriction and contribute to pulmonary hypertension and chronic lung disease. Nitric oxide may maintain the uterus in a quiescent state during pregnancy and may control bladder outflow. It releases insulin from the pancreas, inhibits the release of renin from the kidney and may be involved in the regulation of thyroid hormone production. The adrenal gland contains many nitrergic neurons, too. Neurons containing NOS are present throughout the brain and appear to be relevant in learning, memory, depression, alertness and pain perception. All these known functions of nitric oxide represent therapeutic possibilities.

Induction of nitric oxide synthase in response to cytokines or endotoxins is part of the inflammatory response and could be involved in the vasodilatation, vascular leakage and tissue damage seen in a number of inflammatory conditions such as rheumatoid arthritis, ulcerative colitis and glomerulonephritis. Nitric oxide synthesis is enhanced in patients with septic shock. In such situations drugs to decrease the production or effects of nitric oxide could be helpful. Inhibition of NO with substrate analogues such as L-NMMA has been shown to protect against the cytotoxic effects of cytokines in vitro and to restore the blood pressure of patients with septic shock (31). Selective inhibitors of the inducible isoform of NOS are being developed and selective inhibitors of cerebral nitric oxide synthase have been described.

Nitric oxide can inhibit tumour cell proliferation by both cGMP-dependent and -independent mechanisms. NOS isoforms catalyse the oxidation of arginine to NO plus citrulline; the principal intermediate in the reaction is NG-hydroxy-L-arginine (NOHA) which may play a role as a biological mediator in its own right. Recent experiments using the Caco-2 human colon carcinoma cell line indicate that both NO and NOHA inhibit tumour cell growth (32). Mammalian cell growth requires the production of polyamines from arginine in a cascade of biochemical reactions. NOHA inhibits arginase, the first enzyme in the cascade, and NO may inhibit ornithine decarboxylase (ODC), the second enzyme in the cascade.

There are undoubtedly many more as yet unknown functions of nitric oxide. We can expect plenty more research activity in the field and new agents for the diagnosis, prevention and treatment of many cardiovascular disorders.

References

1. Gruetter CA, Barry BK, McNamara DB, et al. Relaxation of bovine coronary artery and activation of coronary arterial guanylate cyclase by nitric oxide, nitroprusside and a carcinogenic nitrosoamine. J Cyclic Nucleotide Protein Phosphor Res 1979; 5: 211–224.
2. Furchgott RF, Zawadzki JV. The obligatory role of endothelial cells in the relaxation of arterial smooth muscle by acetylcholine. Nature 1980; 288: 373–376.
3. Ignarro LJ, Wood KS, Byrns RE. Pharmacological and biochemical properties of endothelium-derived relaxing factor (EDRF): evidence that EDRF is closely related to nitric oxide (NO) radical (abstract). Circulation 1986; 74 (suppl. II): II–287.
4. Ignarro LJ, Lippton H, Edwards JC, et al. Mechanism of vascular smooth muscle relaxation by organic nitrates, nitrites, nitroprusside and nitric oxide: evidence for the involvement of S-nitrosothiols as active intermediates. J Pharmacol Exp Ther 1981; 218: 739–749.
5. May DC, Popma JJ, Black WH, et al. In vivo induction and reversal of nitroglycerin tolerance in human coronary arteries. N Engl J Med 1987; 317: 805–809.

6. Ignarro LJ. Biological actions and properties of endothelium-derived nitric oxide formed and released from artery and vein. Circ Res 1989; 65: 1–21.
7. Ignarro LJ, Byrns RE, Woods KS. Endothelium-dependent modulation of cGMP levels and intrinsic smooth muscle tone in isolated bovine intrapulmonary artery and vein. Circ Res 1987; 60: 82–92.
8. Rees DD, Palmer RM, Moncada S. Role of endothelium-derived nitric oxide in the regulation of blood pressure. Proc Natl Acad Sci USA 1989; 86: 3375–78.
9. Ayajiki K, Kindermann M, Hecker M, et al. Intracellular pH and tyrosine phosphorylation but not calcium determine shear stress-induced nitric oxide production in native endothelial cells. Circ Res 1996; 78: 750–758.
10. Fleming I, Bauersachs J, Busse R. Calcium-dependent and calcium-independent activation of the endothelial NO synthase. J Vasc Res 1997; 34: 165–174.
11. Ignarro LJ, Harbison RG, Wood KS, Kadowitz PJ. Activation of purified soluble guanylate cyclase by endothelium-derived relaxing factor from intrapulmonary artery and vein; stimulation by acetylcholine, bradykinin and arachidonic acid. J Pharmacol Exp Ther 1986; 237: 893–900.
12. Azuma H, Ishikawa M, Sekizaki S. Endothelium-dependent inhibition of platelet aggregation. Br J Pharmacol 1986; 88: 411–415.
13. Khan MT, Furchgott RF. Similarities of behaviour of nitric oxide (NO) and endothelium-derived relaxing factor in a perfusion cascade bioassay system. Fed Proc 1987; 46: 385 (abstract).
14. Ignarro LJ, Buga GM, Wood KS, Byrns RE, Chaudhuri G. Endothelium-derived relaxing factor produced and released from artery and vein is nitric oxide. Proc Natl Acad Sci USA 1987; 84: 9265–9269.
15. Palmer RM, Ferrige AG, Moncada S. Nitric oxide release accounts for the biological activity of endothelium-derived relaxing factor. Nature 1987; 327: 524–526.
16. Schafer AI, Alexander RW, Handin RI, Inhibition of platelet function by organic nitrate vasodilators. Blood 1980; 55: 649–654.
17. Mellion BT, Ignarro LJ, Ohlstein EH, et al. Evidence for the inhibitory role of guanosine 3',5'-monophosphate in ADP-induced human platelet aggregation in the presence of nitric oxide and related vasodilators. Blood 1981; 57: 946–955.
18. De Rubertis FR and Craven PA. Calcium-independent modulation of cyclic GMP and activation of guanylate cyclase by nitrosamines. Science 1976; 193: 897–889.
19. Ignarro LJ, Degnan JN, Baricos WH, et al. Activation of purified guanylate cyclase by nitric oxide requires heme: comparison of heme-deficient, heme-reconstituted and heme-containing forms of soluble enzyme from bovine lung. Biochim Biophys Acta 1982; 718: 49–59.
20. Wolin MS, Wood KS, Ignarro LJ. Guanylate cyclase from bovine lung in a kinetic analysis of the regulation of the purified soluble enzyme by protoporphyrin IX, heme and nitrosyl-heme. J Biol Chem 1982; 257: 13312–13320.
21. Garthwaite J. Nitric oxide synthesis linked to activation of excitatory neurotransmitter receptors in the brain. In: Nitric oxide from L–arginine; a bioregulatory system (eds. S Moncada, EA Higgs), Excerpta Medica, Elsevier, Amsterdam, 1990, 115–137.
22. Deguchi T, Yoshioka M. L–arginine identified as an endogenous activator for soluble guanylate cyclase from neuroblastoma cells. J Biol Chem 1982; 257: 10147–10152.
23. Bush PA, Aronson WJ, Buga GM, et al. Nitric oxide is a potent relaxant of human and rabbit corpus cavernosum. J Urol 1992; 147: 1650–1655.
24. Tottrup A, Svane D and Forman A. Nitric oxide-mediating NANC inhibition in opossum lower esophageal sphincter. Am J Physiol 1991; 260: G385–G389.
25. Foerstermann U, Kleinert H. Nitric oxide synthase: expression and expressional control of the three isoforms. Naunyn-Schmiedeberg's Arch Pharmacol 1995; 352: 351–364.
26. Vallance P, Moncada S. Role of endogenous nitric oxide in septic shock. New Horizons 1993; 1: 77–87.

27. Moncada S, Palmer RMJ, Higgs EA. Nitric oxide: physiology, pathophysiology and pharmacology. Pharmacol Rev 1991; 43: 109–142.
28. Griffith OW, Stuehr DJ. Nitric oxide synthases: properties and catalytic mechanism. Annu Rev Physiol 1995; 57: 707–736.
29. Bredt DS, Hwang PM, Glatt CE, et al. Cloned and expressed nitric oxide synthase structurally resembles cytochrome P–450 reductase. Nature 1991; 351: 714–718.
30. Vallance P, Moncada S. Nitric oxide: from mediator to medicines. J Roy Coll Phys Lond 1994; 28: 205–218.
31. Petros A, Bennett D, Vallance P. Effect of nitric oxide synthase inhibitors on hypotension in patients with septic shock. Lancet 1991; 338: 1557–1558.
32. Buga GM, Wei LH, Bauer PM, et al. NG–hydroxy–L–arginine and nitric oxide inhibit Caco–2 tumor cell proliferation by distinct mechanisms. Am J Physiol 1998; 275: R1256–R1264.

From an explosive to the endogenous nitrate: Historical, physiological and clinical aspects

T.F. Lüscher, F. Ruschitzka, G. Noll

From an explosive to a remedy

Nitrocompounds have had a long career in science. They are best known in medicine, but their use started in a very different area. Indeed, Alfred Nobel, the founder of the Nobel Prize, produced nitroglycerin as an explosive (Figure 1). His experience with nitroglycerin was initially traumatizing, as his factory exploded in 1864. In 1865 he invented dynamite, which was, and still is, one of the most successful explosives. Indeed, the patent for this invention was the basis for Nobel's success as a businessman and allowed him to start the Nobel Foundation toward the end of his life. The fortune of the Nobel Foundation is still used today for the Nobel prizes in sciences and medicine and is awarded every year.

Figure 1: Portrait of Alfred Nobel (1833–1986).

Nitroglycerin however has had a different career. Its vasoactive effects were obvious. Indeed, workers in nitroglycerin factories experienced typical headaches, particularly on Mondays when they started to work again and were exposed to the substance after a free weekend. The vasodilatory effects of nitroglycerin, however, were only discovered later. The first nitrocompound used in medicine was nitrite of amyl. Indeed, Sir Thomas Lauder Brunton, physician in London, recognized the effectiveness of this compound in patients with angina pectoris and published this observation in The Lancet in 1867 (Figure 2) (1). Interestingly, Brunton only had a vague concept of angina pectoris and its pathogenesis. Also, the mechanisms of action of the nitrocompound were unknown at that point. In fact, Brunton intended to mimic the effects of blood letting – a particularly popular measure at that time – with the drug. After all, even today this approach does not seem completely unreasonable, since the filling pressures of the left ventricle are known to determine myocardial oxygen consumption and in turn ischemia.

ON THE

USE OF NITRITE OF AMYL IN ANGINA PECTORIS.

BY T. LAUDER BRUNTON, B.Sc., M.B.,

SENIOR PRESIDENT OF THE ROYAL MEDICAL SOCIETY, AND RESIDENT PHYSICIAN TO THE CLINICAL WARDS OF THE ROYAL INFIRMARY, EDINBURGH.

Figure 2: The seminal paper, which appeared in the Lancet.

Toward the end of the 19th century nitroglycerin began to be used increasingly to treat patients with angina pectoris. When Nobel himself succumbed to the disease, his physician proposed to prescribe nitroglycerin. But Alfred Nobel absolutely refused to take it; it was inconceivable for him to ingest an explosive to treat his heart disease.

Effects of nitrocompounds in the cardiovascular system

The clinical and hemodynamic effects of nitrovasodilators were then characterized. It became clear that nitroglycerin was a vasodilator, an effect that helped to explain its blood pressure lowering effects in the pulmonary and systemic circulation (Figure 3). Clinically, the reduction in preload is important in coronary artery disease, heart failure and pulmonary edema. Indeed, filling pressure and in turn wall stress is an important determinant of myocardial oxygen consumption that contributes to ischemia in patients with coronary artery disease. Furthermore, the vasodilatory effects in the coronary circulation contribute to an increased myocardial blood supply in these patients. In general, veins are more susceptible to nitrocompounds than are arteries, possibly due to the low basal release of endogenous nitric oxide in their endothelium (Figure 4) (2).

Mechanisms of action of nitrocompounds

But how did nitroglycerin exert its effects? It was Ferid Murad who showed that nitrocompounds increase the formation of a second messenger in the vessel wall, 3',5'-cyclic guanosine monophosphate (cGMP) through the enzyme guanylyl cyclase (Figure 5) (3). The mechanism of activation was clarified by Louis Ignarro, who showed that nitrovasodilators all release nitric oxide (NO), which in turn binds to the heme group, changes its confirmation and activates its catalytic site (4).

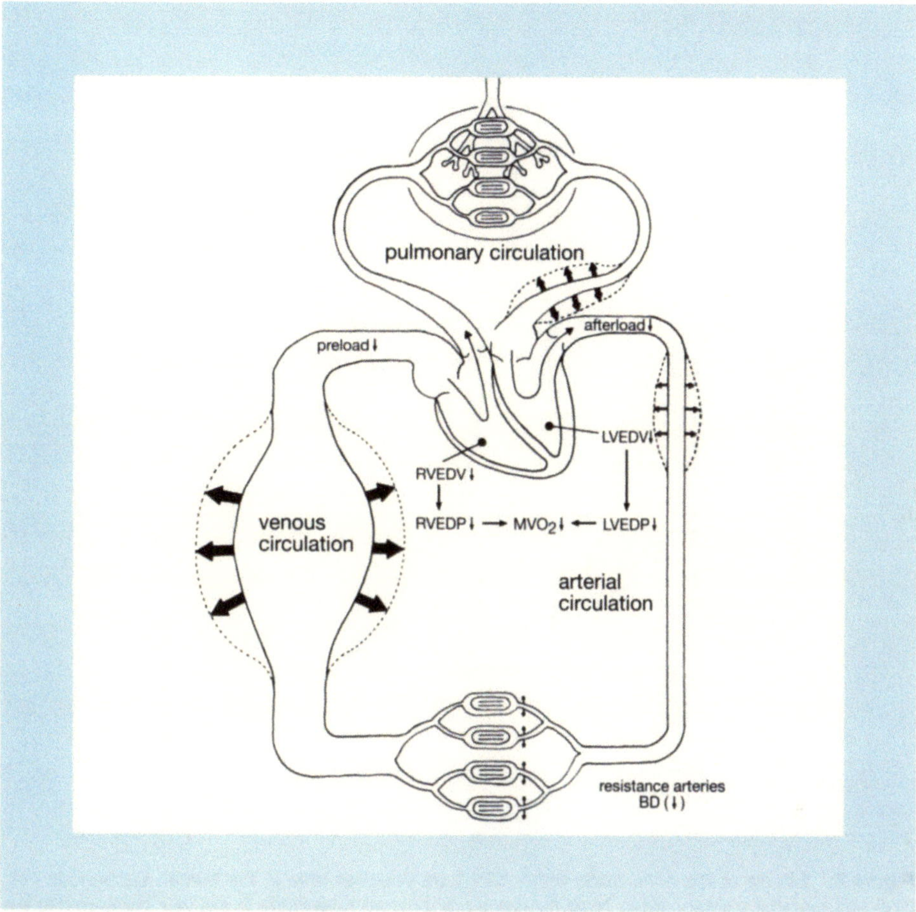

Figure 3: Hemodynamic effects of nitrovasodilators in the circulation.

Science is a history of mistakes

At the same time, Robert Furchgott performed seemingly different experiments in the rabbit aorta; he asked his technician to perform concentration response curves with norepinephrine and acetylcholine, both vasoconstrictors in this preparation (5). For some reason, he mixed up the substances in the middle of the concentration response curve and added acetylcholine on top of an aorta contracted with norepinephrine. He was puzzled to observe a relaxation. The proposal of Robert Furchgott that the response to acetylcholine was mediated by a diffusible substance produced by the endothelial cell, the so-called endothelium-derived relaxing factor (EDRF), opened a completely new and still growing area of research in science and medicine. His proposal in 1985 that EDRF was NO, later confirmed experimentally by the group of Salvador Moncada, closed the loop and led to the discovery of

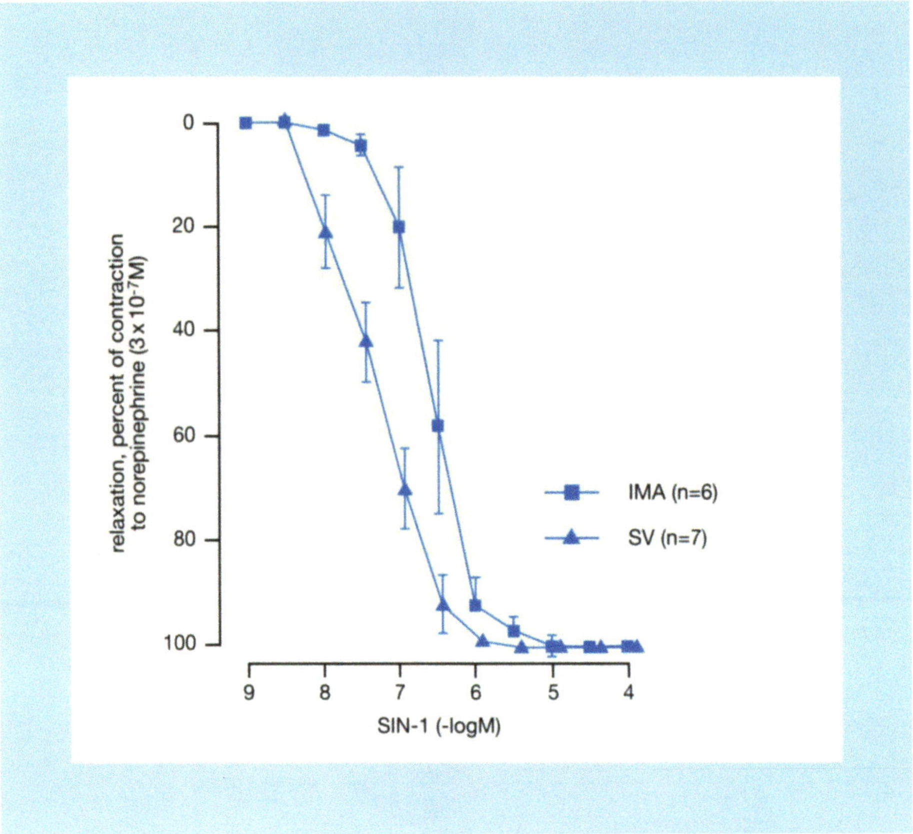

Figure 4: Effects of the nitric oxide donor SIN-1 on vascular tone in the human saphenous vein (SV) and mammary artery (IMA). Note the more pronounced relaxations in the vein compared to the artery.

the endogenous nitrate NO and the pathway leading to its formation from L-arginine via nitric oxide synthase III (NOS III; Figure 5) (6).

Endothelium-derived relaxing factors

Today it is obvious that the endothelium is a source of an array of relaxing factors such as prostaglandins (PGI_2, PGE2), nitric oxide (NO), endothelium-derived hyperpolarizing factor (EDHF) and C-type natriuretic peptide (CNP) (Figure 5) (2, 6).

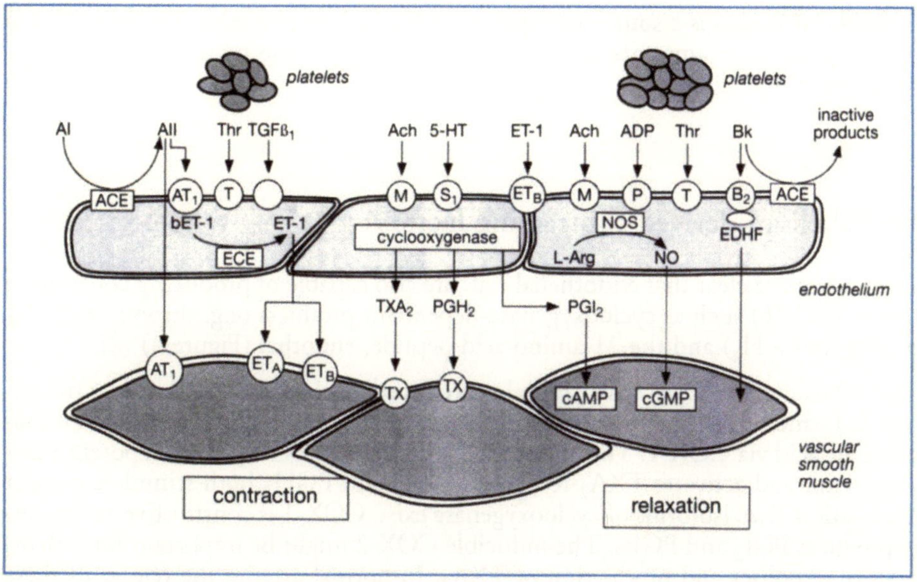

Figure 5: Endothelium-derived vasoactive substances. eNOS = endothelial nitric oxide synthase; NO = nitric oxide; ET = endothelin-1; PGI2 = prostacyclin; TXA2 = thromboxane A2 (8).

PGI_2 is produced from arachidonic acid and converted to PGI2 via an enzyme cascade, i.e. phospholipase A_2, cyclooxygenase (COX-1) and prostacyclin-synthetase. Shear stress, thrombin and hypoxia can stimulate its production. PGI_2 activates adenylyl cyclase and in turn increases cyclic 3',5'-adenosine monophosphate (cAMP) in vascular smooth muscle cells and platelets. PGI_2 exhibits synergistic effects with NO to cause relaxation and platelet inhibition.

NO, the former EDRF of Robert Furchgott, on the other hand, is a free radical with a half-life of a few seconds in biological fluids. It is synthesized from L-arginine via a family of NO-synthases: (1) neuronal (nNOS), (2) endothelial (eNOS) and (3) inducible NOS (iNOS). In endothelial cells eNOS is most dominant, although iNOS may be expressed in the endothelium if activated by cytokines (i.e. interferon, interleukin, tumor necrosis factor, and lipopolysaccharide). Activation of eNOS occurs via specific receptors by acetylcholine, bradykinin, histamine, serotonin, platelet-derived adenine nucleotides and thrombin. Of pathophysiological importance, iNOS may be activated in macrophages, smooth muscle and fibroblasts. NO can freely cross cell membranes and activates soluble guanylate cyclase, which increases the synthesis of cyclic guanosine 3',5'-monophosphate (cGMP) and causes vascular relaxation and inhibition of cell migration. In the vasculature, abluminally released NO inhibits platelet function. In addition, NO modulates cell cycle regulatory proteins to inhibit cell proliferation independent of cGMP.

For proper function of NOS, the cofactor tetrahydrobiopterin (BH4) is required (7). BH4 is formed by GTP cyclohydrolase. With low intracellular concentrations of BH4, NOS also is a source of superoxide (O_2^-). Under these conditions, NO is produced in lesser amounts and inactivated by O_2^- to the toxic product peroxynitrate ($OnOO^-$).

Endothelium-derived contracting factors

Later it became clear that endothelial cells are also capable of producing contracting factors (EDCF) such as cyclooxygenase-dependent products (e.g. thromboxane A_2, prostaglandin H_2) and the 21 amino acid peptide, endothel (Figure 4) (8).

Thromboxane A_2 (TXA_2), which is also produced in platelets stimulated by thrombin, is formed from the precursor prostaglandin H_2 (PGH_2) converted from arachidonic acid via COX-1. Thromboxane A_2 synthetase. PGH_2 is also a potent vasoconstrictor and activates TXA_2 receptors. TXA_2 and PGH_2 both stimulate platelet aggregation. Two isoforms of cyclooxygenase exist, COX-1 is constitutive and mainly produces PGI_2 and PGE_2. The inducible COX-2 might be important for inflammatory responses and production of TXA_2. In normal arteries the release of these factors is limited. In veins, however, the endothelium releases TXA_2 and PGH_2 in response to acetylcholine, histamine and other autacoids. TXA_2 and PGH_2 antagonize the vasodilatory and anti-aggregatory effects of NO. The COX pathway is also a source of O_2^-, as are NAD(P)H oxidases and nitric oxide synthase itself (in the presence of low levels of tetrahydrobiopterin). The superoxide anion itself is not a constricting factor but evokes vascular contraction only indirectly by breakdown of NO.

Endothelin is a 21 amino-acid peptide(9). Three isoforms of endothelin are known ET-1, ET-2, ET-3. ET-1 is the main product of the endothelium if stimulated by thrombin, angiotensin II, vasopressin, TGF-1, cytokines and others. Big ET-1, a 38 amino acid peptide, is an end product of preproendothelin and is converted to ET-1 via endothelin-converting enzymes (ECE-1 and ECE-2) (10). Most interestingly in this context, endothelin production is inhibited by NO via a cGMP-dependent pathway. Accordingly the inhibitor of NO production, L-NMMA, increases, while nitroglycerin, as well as the nitrovasodilator SIN-1, reduces vascular endothelin production.

ET-1 induces vasoconstriction in vitro and in vivo via two receptors, ET_A and ET_B, on vascular smooth muscle cells (Figure 5) (11). Endothelial ET_B receptors release PGI_2 and NO. This explains why ET-1 causes transient vasodilation followed by long-lasting vasoconstriction. Apart from its direct vascular effects, ET-1 potentiates the effects of other vasoconstrictor hormones (i.e. serotonin, norepinephrine) and of growth factors (PDGF) (12).

The endothelium in cardiovascular disease

Aging is a main determinant for the development of cardiovascular disease. In arteries exposed to high pressure, aging is associated with reduced NO production. Since NOS expression is actually increased, as is the production of free oxygen-derived free radicals, aging seems to primarily affect the breakdown of NO by O_2^-. ET-1 expression also increases with age.

In hypercholesterolemia endothelium-dependent relaxations are impaired (13, 14). These effects are mainly due to oxidized low density lipoproteins (ox-LDL) (Figure 6) (15) which activate scavenger-like receptors. Specific effects on Gi-proteins impair signal transduction, and diminished intracellular BH4 and L-arginine levels decrease NOS function and activity. At this stage of the disease process NOS expression is normal or increased. NO, however, is inactivated by O_2 produced by dysfunctional NOS and endothelial NAD(P)H oxidases. In fully developed atherosclerosis, NOS is downregulated and NO production severely impaired.

In hypertension, endothelial dysfunction, i.e. impaired relaxations and/or reduced basal NO release, occurs in most experimental models (Figure 7) (16). The mechanisms involved differ, however. In the SHR, NOS dysfunction with production of superoxide occurs early. In addition, endothelium-derived TXA_2/PGH_2 is released and reduces endothelium-dependent relaxations. ET plays a minor role in this model. In salt-sensitive hypertension, NO production is reduced and ET-1 is

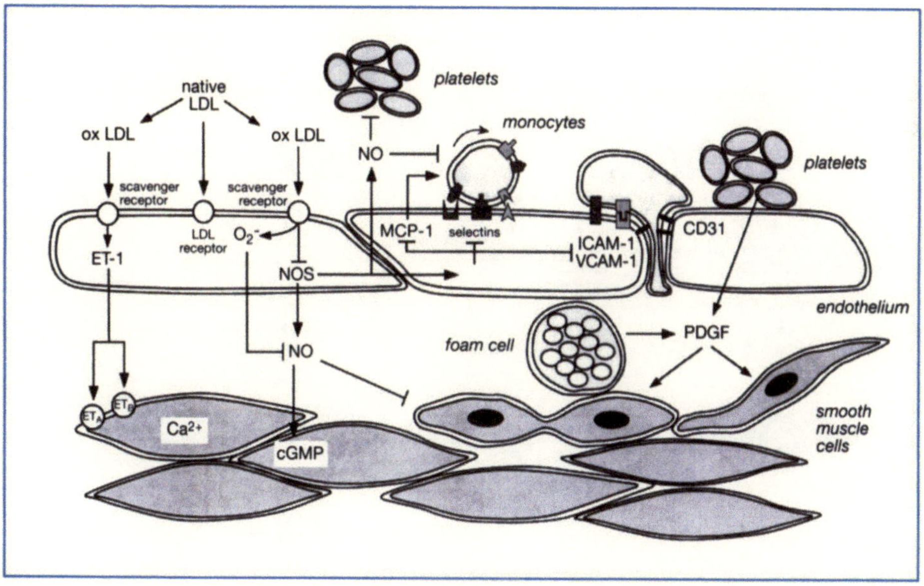

Figure 6: Role of endothelial factors in atherosclerosis. eNOS = endothelial nitric oxide synthase; MCP-1 = monocyte chemoattractant protein 1; LDL = low density lipoprotein; NO = nitric oxide; ET = endothelin-1; PDGF = platelet-derived growth factor (14).

increased. ET receptor blockade normalizes NO production and the vascular structure in this model. Chronic angiotensin infusion increases vascular ET content and causes vascular hypertrophy, an event that is prevented by ET blockade (17).

In hypertensive patients, the response to acetylcholine has been shown to be blunted in most, but not all, studies (18). Also, the vasoconstrictor effect of the NO inhibitor LNMMA is reduced, indicating a reduced NO release. Direct measurements of NO plasma levels also revealed reduced levels (19).

Many of the events occuring in atherosclerosis could be related to endothelium-dependent mechanisms (14). Atherosclerosis is characterized by three major abnormalities, i.e. (1) abnormal vasomotion with predominant vasoconstriction, (2) structural alterations of the vascular wall (with outward and later inward remodeling) and (3) thrombus formation. Many of these features are related to endothelial dysfunction. The abnormal vasomotion can be explained by reduced NO bioavailability or production, and increased ET-1 production in the atherosclerotic vessel wall. Structural alterations are related to the invasion of monocytes and other white blood cells (i.e. lymphocytes, mast cells) and in turn the formation of foam cells, which contain oxidized LDL. The invasion of monocytes is governed by chemoattractant proteins and adhesion molecules (Figure 6), which are also regulated by NO and oxygen-derived free radicals via the transcription factor Nf kappa b. In addition, the atherosclerotic plaque contains smooth muscle cells, which are activa-

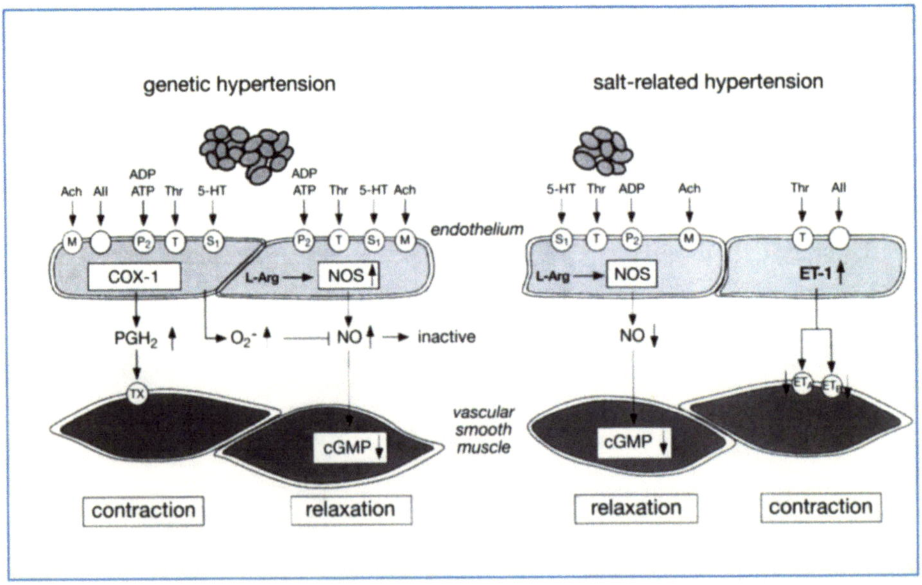

Figure 7: Endothelial dysfunction in hypertension. eNOS = endothelial nitric oxide synthase; O_2^- = superoxide; NO = nitric oxide; ET = endothelin-1; PGH_2 = endoperoxide; TXA_2 = thromboxane A_2 TXA_2 = thromboxane A_2.

ted by growth factors. ET markedly potentiates these responses in human smooth muscle cells without its own genuine proliferative effects. NO, on the other hand, acts as an inhibitor of migration and proliferation (Figure 6). Finally, the endothelium is an important antithrombotic barrier and via the release of NO, and PGI_2 is an important inhibitor of platelet adhesion and aggregation. The reduced NO release in atherosclerosis could contribute to the prothrombotic state occurring under these conditions (Figure 6).

Acute coronary syndromes markedly increase the risk of death and myocardial infarction. They are initiated either by plaque rupture or endothelial erosion (20). Both events lead to the exposure of subendothelial structures with massive thrombus formation. Plaque rupture occurs typically in lipid-rich plaques with monocyte and lymphocyte infiltrates in the shoulder region. Hence, uptake of lipids and monocytes is a primary event in the formation of unstable plaques. Metalloproteases (MMP-1, MMP-9) produced by these white blood cells digest collagen and further destabilize the plaque. Abnormal vasomotion might act as a triggering event for plaque rupture, explaining the known association of acute coronary syndromes with heavy exercise. Indeed, vascular tissue ET-1 content is increased in patients with unstable angina or angina after acute myocardial infarction. Furthermore, activation of inflammatory mediators in unstable angina is associated with "endothelial stunning". Endothelium-dependent vasodilation to acetylcholine is markedly blunted under these conditions. A reduced endothelium-dependent vasodilation of the coronary circulation may be important prognostically. Indeed, local blood flow in a coronary segment is an important determinant of thrombus formation in patients with plaque rupture and hence of the risk of a myocardial infarction. In patients with coronary artery disease, impaired responses of the coronary circulation to acetylcholine are associated with a poor prognosis. Endothelial erosion is less well understood, but may be related to apoptotic mechanisms together with mechanical factors.

Clinical role of nitrates today

One obvious therapeutic measure in states with a reduced production of a given mediator would be its pharmacological substitution. Nitrate therapy would therefore seem to be the most logical approach. Indeed, nitrovasodilators relax blood vessels in the presence and absence of the endothelium. In fact, in normal blood vessels, removal of the endothelium even enhances the effects of NO donors (Figure 3). In patients with coronary artery disease, particularly at later stages of the disease process, the response of the coronary circulation to nitroglycerin may be reduced as well. Nevertheless, the paradoxical vasoconstriction of coronary arteries occurring during exercise in patients with coronary artery disease is reversed by nitroglycerin. This may at least in part explain the anti-anginal effects of the compound in stable angina. The route of application is commonly oral. Comparative studies using coronary angiography suggest that the vasodilator effects of the spray are comparable to intracoronary application and superior to the classical capsules (Figure 8) (21).

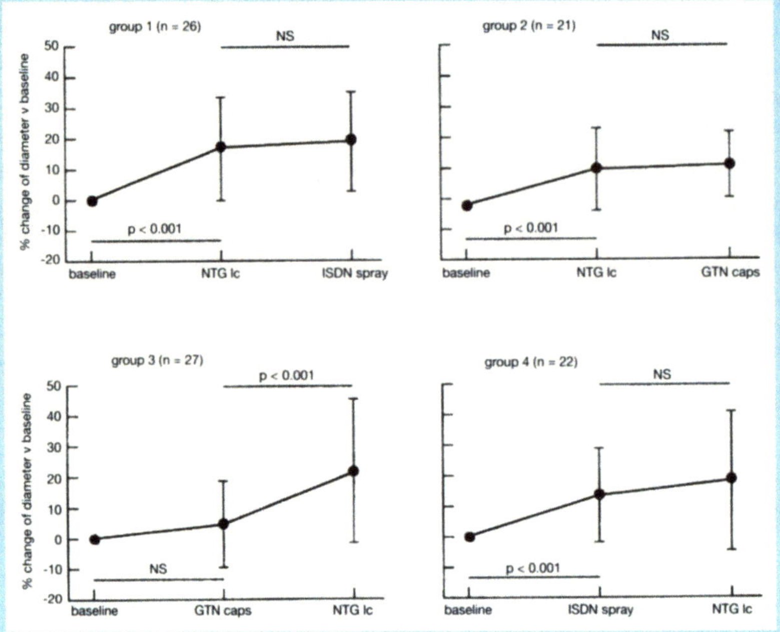

Figure 8: Effects of nitroglycerin given either intracoronary (NTG i.c.), as capsules (GTN caps) orally or spray (ISDN spray) applied orally on coronary diameter (21).

Recent findings showed that nitrates only in part mimic the effects of the endogenous nitrovasodilator system. Indeed, systemic application of nitrates has unfavorable hemodynamic effects, in particular activation of the sympathetic nervous system, not shared by the endogenous nitiric oxide system, which is primarily activated locally within the vessel wall. Furthermore, in rabbits with nitrate tolerance vascular superoxide production, as well as endothelin production, is enhanced (22). These latter effects not only contribute to the systemic and vascular aspects of nitrate tolerance, but possibly may also alter vascular function via increased levels of peroxynitrite (formed after inactivation of NO by superoxide) and ET. Clinically, however, at least in stable angina pectoris, nitrate tolerance seems to be less of a problem, with intermittent therapy leaving a drug-free therapeutic window overnight.

In heart failure, nitrates are useful both in acute cases and during chronic treatment. Nitrates reduce filling pressures in patients with heart failure and in turn ameliorate symptoms rapidly. The V-HeFT trial showed that triglyceril nitrate, in combination

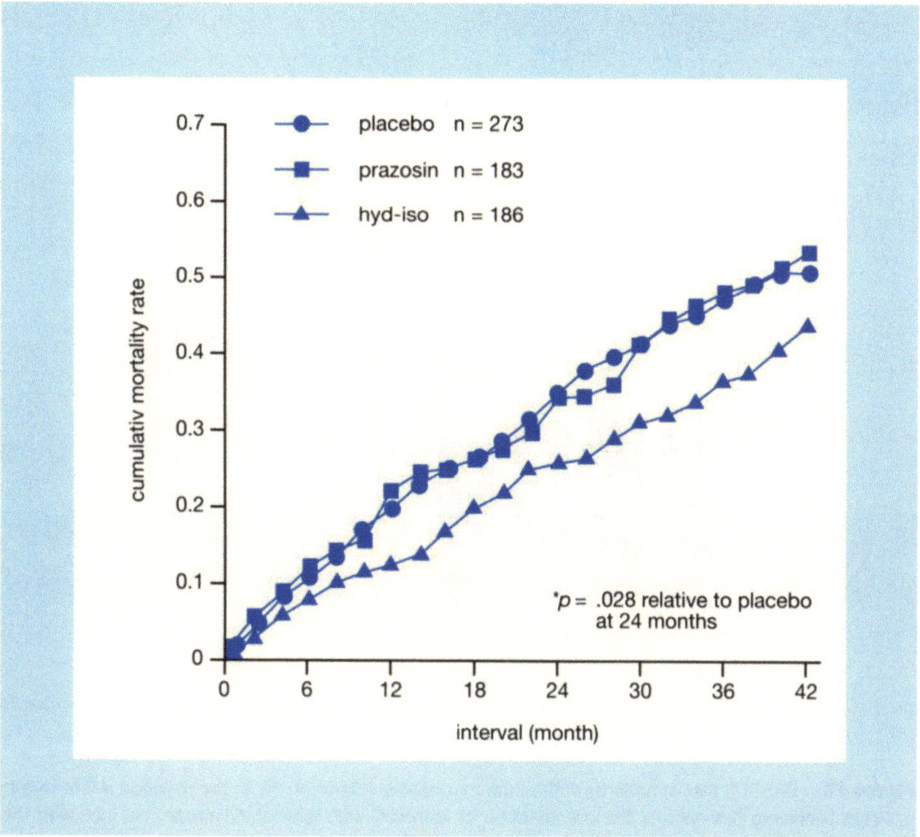

Figure 9: Results of the V-HeFT-1 trial in patients with heart failure. Note the reduced cardiovascular event rate in patients receiving isosorbiddinitrate plus hydralazine (hyd-iso), but not in those receiving the alpha blocker prazosin (23).

with hydralazine, reduced mortality in patients with heart failure (Figure 9) (23). Unfortunately, a trial investigating monotherapy with a nitrocompound versus placebo has not been performed.

Given the evidence in heart failure, nitrates also seemed excellent drugs in patients after myocardial infarction. Indeed, early studies looking at the remodeling process suggested that nitrates reduce dilatation of the left ventricle after an acute myocardial infarction. The GISSI-3 trial tested the hypothesis in a large group of patients with acute myocardial infarction. Although a positive trend was noted, the difference was not significant (Figure 10) (24). It should be emphasized that a substantial number of the patients in the placebo group also received nitrates, limiting the conclusions of this trial. Furthermore, there was some indication that the nitrates were particularly effective in patients also receiving ACE inhibitors.

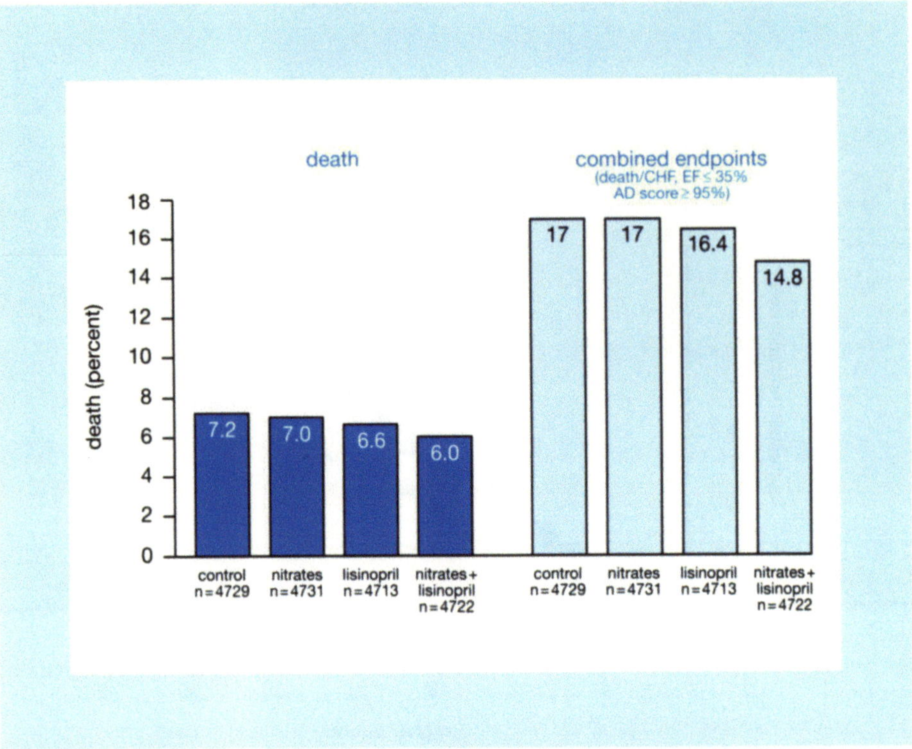

Figure 10: GISSI-3 trial in patients with acute myocardial infarction. Note the reduced event rate in patients receiving lisinopril or the combination of lisinopril with isosorbiddinitrate, but not with the nitrate alone (24).

Gene therapy

It is likely, however, that many of these therapeutic interventions remain ineffective, either at later stages of the disease process when NOS expression is lost, or in patients with mutations of the NOS gene. If so, it would make much more sense to substitute the mutated or underexpressed gene by a normal variant, i.e. to use somatic gene therapy with a vector containing the normal eNOS gene to treat these patients.

Acknowledgments

Personal research of the authors has been supported by the Swiss National Research Foundation (grant Nr. 32-51069.97/1) and the Swiss Heart Foundation.

References

1. Brunton L. On the use of amyl in angina pectoris. Lancet. 1867.
2. Lüscher TF, Vanhoutte PM. The Endothelium: Modulator of Cardiovascular Function. Boca Ranton: CRC Press, 1990.
3. Murad F. Cyclic guanosine monophosphate as a mediator of vasodilation. J Clin Invest. 1986; 78: 1–5.
4. Ignarro LJ, Buga GM, Wood KS, et al. Endothelium-derived relaxing factor produced and released from artery and vein is nitric oxide. Proc Natl Acad Sci U.S.A. 1987; 84: 9265–9269.
5. Furchgott RF, Zawadzki JV. The obligatory role of endothelial cells in the relaxation of arterial smooth muscle by acetylcholine. Nature 1980; 299: 373–376.
6. Palmer RM, Ashton DS, Moncada S. Vascular endothelial cells synthesize nitric oxide from L-arginine. Nature 1988; 333: 664–666.
7. Cosentino F, Patton S, d'Uscio LV, et al. Tetrahydrobiopterin alters superoxide and nitric oxide release in prehypertensive rats. J Clin Invest 1998; 101: 1530–1537.
8. Lüscher TF, Boulanger CM, Dohi Y, Yang ZH. Endothelium-derived contracting factors. Hypertension 1992; 19: 117–130.
9. Yanagisawa M, Kurihara H, Kimura S, et al. A novel potent vasoconstrictor peptide produced by vascular endothelial cells. Nature. 1988; 332: 411–415.
10. Xu D, Emoto N, Giaid A, et al. ECE-1: a membrane-bound metalloprotease that catalyzes the proteolytic activation of big endothelin-1. Cell 1994; 78: 473–485.
11. Sakurai T, Yanagisawa M, Takuwa Y, et al. Cloning of a cDNA encoding a non-isopeptide-selective subtype of the endothelin receptor [see comments]. Nature 1990; 348: 732–735.
12. Yang ZH, Richard V, von Segesser L, et al. Threshold concentrations of endothelin-1 potentiate contractions to norepinephrine and serotonin in human arteries. A new mechanism of vasospasm? Circulation 1990; 82: 188–195.
13. Anderson TJ, Meredith IT, Yeung AC, et al. The effect of cholesterol-lowering and antioxidant therapy on endothelium-dependent coronary vasomotion. N Engl J Med 1995; 332: 488–493.
14. Lüscher TF, Tanner FC, Tschudi MR. Endothelial dysfunction in coronary artery disease. Annu Rev Med. 1993; 44: 395–418.
15. Tanner FC, Noll G, Boulanger CM, Lüscher TF. Oxidized low density lipoproteins inhibit relaxations of porcine coronary arteries. Role of scavenger receptor and endothelium-derived nitric oxide. Circulation 1991; 83: 2012–2020.
16. Lüscher TF, Vanhoutte PM. Endothelium-dependent contractions to acetylcholine in the aorta of the spontaneously hypertensive rat. Hypertension 1986; 8: 344–348.
17. Moreau P, d'Uscio LV, Shaw S, et al. Angiotensin II increases tissue endothelin and induces vascular hypertrophy: reversal by ET(A)-receptor antagonist. Circulation 1997; 96: 1593–1597.
18. Taddei S, Virdis A, Mattei P. Vasodilatation to acetylcholine in primary and secondary forms of hypertension. Hypertension 1993; 21: 929–933.
19. Forte P, Copland M, Smith LM, et al. Basal nitric oxide synthesis in essential hypertension [see comments]. Lancet 1997; 349: 837–842.
20. Ross R. Atherosclerosis – an inflammatory disease [see comments]. N Engl J Med 1999; 340: 115–126.
21. Pfister M, Seiler C, Fleisch M, et al. Nitrate induced coronary vasodilatation: differential effects of sublingual application by capsule or spray. Heart 1998; 80: 365–369.
22. Münzel T, Harrison DG. Increased Superoxide in Heart Failure: A Biochemical Baroreflex Gone Awry. Circulation 1999; 100: 216–218.
23. Cohn JN, Archibald DG, Ziesche S, et al. Effect of vasodilator therapy on mortality in chronic congestive heart failure. Results of a Veterans Administration Cooperative Study. N Engl J Med 1986; 314: 1547–1552.

24. GISSI-3. GISSI-3: effects of lisinopril and transdermal glyceryl trinitrate singly and together on 6-week mortality and ventricular function after acute myocardial infarction. Lancet 1994; 343: 1115–1122.

The role of nitrates in the therapy of congestive heart failure

Ch.J.F. Holubarsch

Introduction

Congestive heart failure is a syndrome of specific clinical symptoms and may develop due to a variety of diseases of the heart: (1) coronary artery disease, (2) idiopathic dilated cardiomyopathy, (3) hypertensive heart disease, and (4) valvular heart disease. The mortality of patients suffering from congestive heart failure is extremely high and comparable to malignant tumors: About half of the patients die of sudden cardiac death, the other half die of progression of heart failure (1).

Recommendations for chronic pharmacologic therapy of congestive heart failure are based on large-scale mortality trials ("evidence-based medicine"). This type of study exists for ACE inhibitors, ß-adrenoceptor blockers, digitalis, and – very recently – for spironolactone.

There are a good number of mortality trials with ACE inhibitors, both in patients with congestive heart failure and those with myocardial infarction. All of these studies show significant improvement of prognosis in these patients (2–6). This type of medical therapy also improves symptoms, and its effect is dose-dependent (7).

More recently, three mortality trials with the ß-adrenoceptor blockers, Carvedilol (8), Bisoprolol (9) and Metoprolol-Zok (10), showed a prolongation of life in patients with congestive heart failure. However, in this therapeutic strategy it is important to start with a very low dosage, to titrate the medication very slowly and to expect objective, and especially subjective, improvement not before three to six months.

Regarding positive inotropic compounds acting via cyclic AMP, negative effects on survival of heart failure patients have been shown: Xamoterol (11), Milrinone (12), and Dobutamin (13) may decrease the survival rate in heart failure patients. Considering these facts it was very important to test the effect of digitalis on mortality. Digitalis is the oldest compound for treatment of congestive heart failure, and has been used in this indication for more than 200 years. Digoxin has neither beneficial and nor detrimental effects regarding survival of heart failure patients (14); however, digitalis was shown to reduce significantly the need for rehospitalization (14). Therefore, digitalis is thought to be efficacious in heart failure but without a significant influence on prognosis. Its use is thereby limited to severly ill patients for symptomatic reasons only.

For loop diuretics and thiazides, no mortality trials have been conducted so far, although they are without doubt essential for treatment of severe heart failure. However, quite recently, spironolactone – an aldosterone-antagonist – has been investigated regarding survival rate of heart failure patients: Spironolactone significantly reduced the mortality rate when given in low dosages (25mg) in addition to standard therapy (15). At the moment it is not clear of whether this beneficial effect is brought about through its diuretic properties or specific influence on collagen production of fibrocytes (16).

Nitrates in heart failure and nitrate tolerance

Nitrates have been used successfully worldwide for many decades in cardiovascular emergency cases, i. e., acute myocardial infarction, instable angina, and severe heart failure with lung congestion or oedema. In those situations, the beneficial effect of nitrates is explained by nitrate-induced vasodilation of epicardial coronary arteries and the venous capacity vessel system. The latter effect results in a preload reduction and improvement of lung congestion symptoms. In those situations, the vasodilation of the arterial peripheral vessels is of less importance.

The acutely beneficial effects of nitrates in the above described emergency care are clearly evident and not at all in doubt. In contrast, long-term therapy with nitrates has been judged critically because of development of the well-known nitrate tolerance (Figure 1). During a 24- or 48-hour-period, nitrate tolerance develops with regard to central hemodynamics (17, 18). Because of such a nitrate tolerance development, long-term treatment with nitrates has been questioned.

However, the underlying mechanisms of nitrate tolerance seem to have been detected recently:

(1) Development of nitrate tolerance depends on the daily dosing schedule: In a semantic study, Parker and coworkers clearly demonstrated that intermittent dosing of nitrates can almost completely prevent nitrate tolerance (Figure 1) (17).
(2) Newest findings from basic research on nitric oxide interpret nitrate tolerance in the light of "oxidative stress". An increased production of oxygen radicals inactivates endogenous and exogenous NO molecules and inhibits the beneficial vasodilatory properties of NO (19).

Acetylcholine – applied at low dosages intracoronarily – leads to increased release of NO from the endothelium and vasodilation in healthy people. In patients with endothelial dysfunction, a "precursor" of coronary artery disease, such a diagnostic intervention induces vasoconstriction, because NO reacts with superoxide anions to peroxynitrate so that the direct vasoconstrictive effect of acetylcholine on vascular smooth muscle cells dominates (19, 20). This endothelial dysfunction can be normalized, or at least positively influenced, by either acute intravenous application of vitamin C or chronic treatment with ACE inhibitors or angiotensin-receptor blockers (21, 22, 23). This means that reduction of oxydative stress prevents inactivation of NO, which can therefore exert its beneficial effect at the smooth muscle cell, i.e., vasodilation (19).

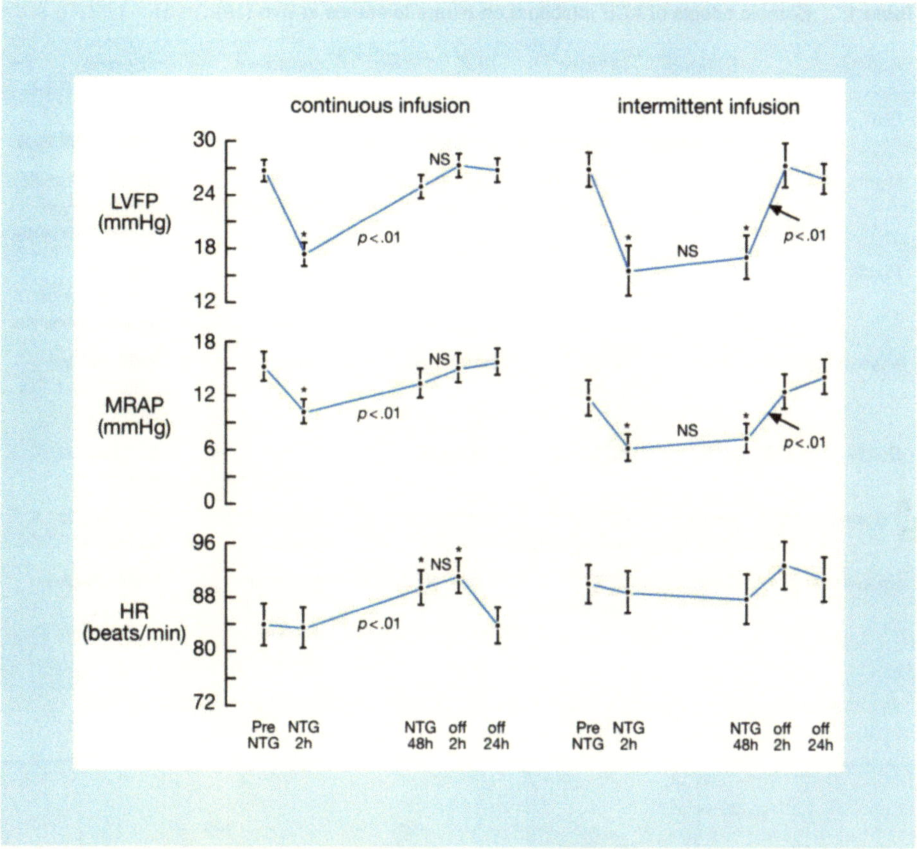

Figure 1: Continuous infusion of nitroglycerin (left) and intermittent infusion of nitroglycerin (right). The effects of nitroglycerin are completely maintained, even after 48 hours in the case of intermittent therapy, whereas nitrate tolerance is evident with continuous infusion (17).

If nitrate tolerance has the same biological basis as endothelial dysfunction, nitrate tolerance could be abolished, or at least attenuated, by concomitant ACE inhibition. Indeed, in animal experiments it has been shown that nitrate tolerance is prevented by previous application of the ACE inhibitor enalapril or AT1 receptor blockade (24, 25). But which impact do these results have on clinical practice?

In a number of small studies this question has been investigated: the seven available studies are listed in Table 1 (26–32): the first four studies clearly show that nitrate tolerance is almost completely prevented by ACE inhibition. This effect seems to be independent of the type of ACE inhibitor (Figure 2) (26). In the next two studies shown in Table 1, nitrate tolerance could not be prevented (30, 31). However, this failure can be explained either by too low ACE inhibitor dosages or too late start of ACE inhibition. Again, a very recent study showed that transdermal NTG on top of ACE inhibition can prolong exercise time, as well as increase echocardiographic fractional shortening (32).

Table 1: Chronic effects of ACE inhibitors on nitrate tolerance in vivo (19).

Authors	Patients	Nitrates	ACE inhibition	endpoints	outcome
Katz et al. (26)	healthy volunteers	NTG transdermal	Captopril 75 mg Enalapril 10 mg	vasodilation	prevention of nitrate tolerance
Mehra et al. (27)	CHF	ISDN oral	Captopril 90 mg	central hemodynamics	almost complete prevention of nitrate tolerance
Pizulli et al. (28)	CAD	NTG i.v.	Captopril 50 mg	central hemodynamics	complete prevention of nitrate tolerance
Muiesan et al. (29)	CAD	NTG transdermal	Benazepril 20 mg	exercise tolerance	combination better than NTG alone
Dodak et al. (30)	CHF	NTG i.v.	Captopril	central hemodynamics	no effect on tolerance
Parker et al. (31)	healthy volunteers	NTG transdermal	Benazepril	arterial pressure	no effect on tolerance
Elkayam et al. (32)	CHF	NTG transdermal	Captopril	exercise tolerance, fractional shortening	NTG effective

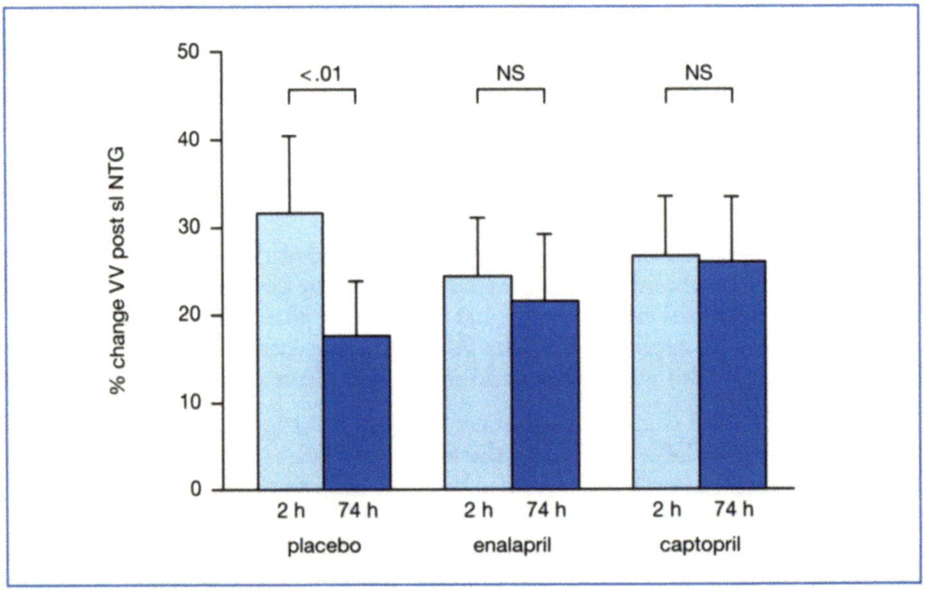

Figure 2: Loss of effectiveness of nitroglycerin when applied over 72 hours (left, placebo). Effectiveness of nitroglycerin is maintained over 72 hours by enalapril or captopril (right) (26) (% change VV post se NTG: percent change of venous volume after 0.4 mg sublingual nitroglycerin).

Mortality trials with nitrates

Two mortality trials have been conducted investigating nitrates (33, 34). In the V-HeFT-I-study, the combination of hydralazine and isosorbide dinitrate significantly reduced mortality of patients with congestive heart failure as compared to placebo or prazosin, an alpha-receptor blocker (33) (Figure 3). In the V-HeFT-II-study, the same combination was studied versus the ACE inhibitor enalapril (34). Enalapril was significantly more effective than the combination of hydralazine and isosorbide dinitrate regarding reduction of mortality.

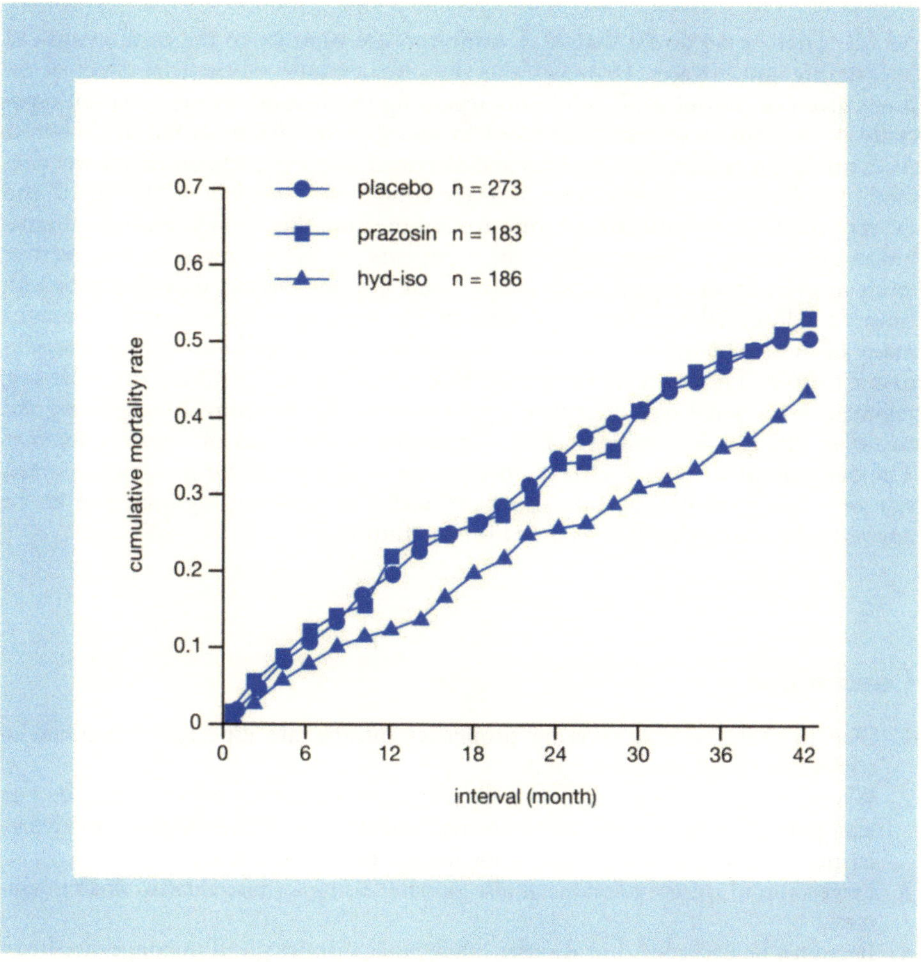

Figure 3: Cumulative mortality of heart failure patients. The combination of hydralazine and isosorbide dinitrate significantly reduces mortality as compared to placebo or prazosin (33).

Very often, two – from our viewpoint – wrong conclusions are drawn from these two studies: (1) In the V-HeFT-I-study only hydralazine is effective, because the continuous dosing of nitrates must lead to nitrate tolerance. (2) ACE inhibition therapy is significantly superior to the combination of hydralazine and isosorbide dinitrate, concluding that there is neither a place for hydralazine nor for nitrates in the medical therapy of congestive heart failure.

Ad (1): Development of nitrate tolerance in the V-HeFT-study is unlikely because of two newer important studies: Already in 1991, Bauer and Fung (35) showed in an animal experimental study that hydralazine prevents nitrate tolerance by the same mechanism as ACE inhibitors. These data could be proven more recently by Münzel et al. (36): Hydralazine prevents activation of membrane-bound NADH oxidase. Thereby the co-medication of hydralazine guarantees chronic effectiveness of nitrates.

Ad (2): There is no doubt that ACE inhibitors are superior to the combination of hydralazine and nitrates. However, this does not exclude a beneficial effect of nitrates given on top of ACE inhibitors regarding the hemodynamic situation especially in the more severe and advanced forms of congestive heart failure. Whereas ACE inhibitors reduce both preload and afterload, nitrates preferentially lower preload (37). Patients with moderate to severe congestive heart failure NYHA III and IV very often show quite low systolic arterial pressure due to high dose ACE-inhibition and diuretics. Nevertheless, these patients are symptomatic at low exercise levels or even at rest regarding dyspnea. These are the patients that clearly benefit from an additional application of nitrates for hemodynamic reasons. Moreover, many of these patients do not tolerate additional ß-receptor blockade for hemodynamic reasons. The co-medication with ACE inhibitors and the intermittent dosing regimen in the sense of a once daily application in the morning – considering the circadian changes in hemodynamics – guarantee prevention of nitrate tolerance. The question of whether such a co-medication – ACE inhibitor plus nitrate – acts not only symptomatically, but also beneficially influences prognosis, has to be answered by a mortality trial planned for the future.

Conclusions

1. Due to their preload-lowering properties nitrates are especially beneficial in advanced congestive heart failure.
2. When nitrates are applied chronically, the expected nitrate tolerance can be – at least partially – circumvented by co-medication with ACE inhibitors (or hydralazine) according to the present knowledge from the literature.
3. Prevention of nitrate tolerance is also possible using an intermittent dosing regimen.
4. It cannot be excluded, but is rather likely, that nitrates given in congestive heart failure have a beneficial effect on prognosis provided they are applied in conjugation with ACE inhibitors.

5. **Patients with severe forms and stages of heart failure do need compounds that improve symptoms, like loop diuretics and nitrates, in addition to those important pharmacological interventions that improve survival.**

References

1. Smith ThW, Kelly RA, Stevenson LW, et al. Management of Heart Failure. In Heart Disease, a Textbook of Cardiovascular Medicine (ed. E Braunwald), 5th Edition, W.B. Saunders Company, Philadelphia-London-Toronto-Montreal-Sydney-Tokyo, 1997, 492–514.
2. CONSENSUS Trial Study Group: Effects of enalapril on mortality in severe congestive heart failure. N Engl J Med 1987; 316: 1429–1436.
3. SOLVD Investigators. Effect of enalapril on survival in patients with reduced left ventricular ejection fractions and congestive heart failure. N Engl J Med 1991; 325: 293–302.
4. SOLVD Investigators. Effect of enalapril on mortality and the development of heart failure in asymptomatic patients with reduced left ventricular ejection fractions. N Engl J Med 1992; 327: 685–691.
5. Pfeffer MA, Braunwald E, Moye LA, et al: Effect of captopril on mortality and morbidity in patients with left ventricular dysfunction after myocardial infarction. N Engl J Med 1992; 327: 669–677.
6. The AIRE study investigators. Effect of ramipril on mortality and morbidity of survivors of acute myocardial infarction with clinical evidence of heart failure. Lancet 1993; 342: 821–828.
7. Packer M. Results of the ATLAS-study, presented during the 47th Annual Scientific Session of the American College of Cardiology, Atlanta, March 29–April 1, 1998.
8. Packer M, Bristow MR, Cohn JN, et al. The effect of carvedilol on morbidity and mortality in patients with chronic heart failure. N Engl J Med 1996; 334: 1349–1355.
9. Dargie HJ, Lechat P for the CIBIS-II investigators and committee: The cardiac insufficiency bisoprolol study II (CIBIS-II). Lancet 1999; 353: 9–13.
10. Hjalmarson A, Goldstein S on behalf of the MERIT-HF study group: mortality results of the metropolol CR/Zok randomized intervention trial in heart failure, presented during the 48th Annual Scientific Session of the American College of Cardiology, New Orleans, March 7–10, 1999.
11. The xamoterol in severe heart failure study group: Xamoterol in severe heart failure. Lancet 1990; 336: 1–6.
12. Packer M, Arver JR, Rodeheffer RJ, et al. Effects of oral milrinone on mortality in severe chronic heart failure. N Engl J Med 1991; 325: 1468–1475.
13. Dies K, Krell M, Whittlow P, et al. Intermittent dobutamine in ambulatory outpatients with chronic cardiac failure. Circulation 1986; 74 (suppl II): 38.
14. The Digitalis Investigation Group: The effect of digoxin on mortality and morbidity in patients with heart failure. N Engl J Med 1997; 336: 525–533.
15. Pitt B, Zannad F, Remme WJ, et al. The effect of spironolactone on morbidity and mortality in patients with severe heart failure. N Engl J Med 1999; 341: 709–717.
16. Brilla CG, Matsubara LS, Weber KT. Anti-aldosterone treatment and the prevention of myocardial fibrosis in primary and secondary hyperaldosteronism. J Mol Cell Cardiol 1993; 25: 563–575.
17. Packer, M, Lee W, Kessler, et al. Prevention and reversal of nitrate tolerance in patients with congestive heart failure. N Engl J Med 1987; 317: 799–804.

18. Roth A, Kulick D, Freidenberger L, et al. Early tolerance to hemodynamic effects of high dose transdermal nitroglycerin in responders with severe chronic heart failure. J Am Coll Cardiol 1987; 9: 858–864.
19. Münzel T, Heitzer T, Brockhoff C. Neurohormoral activation and nitrate tolerance: implications for concomitant therapy with angiotensin-converting enzyme inhibitors or angiotensin receptor blockers. Am J Cardiol 1998; 81 (suppl. 1A): 30A–40A.
20. Drexler H, Zeiher AM, Bassenge E, et al. (eds.): Endothelial mechanisms of vasomotor control, Steinkopff, Darmstadt and Springer, New York 1991.
21. Solzbach U, Hornig B, Jeserich M, et al. Vitamin C improves endothelial dysfunction of epicardial coronary arteries in hypertensive patients. Circulation 1997; 96: 1513-1519.
22. Mancini GBJ, Henry GC, Macaya C, et al. Angiotensin-converting enzyme inhibition with quinapril improves endothelial vasomotor dysfunction in patients with coronary artery disease the TREND strudy. Circulation 1996; 94: 258–265.
23. Rajagopalan S, Kurz S, Münzel T, et al. Angiotensin II-mediated hypertension in the rat increases vascular superoxide production via membrane NADH/NADPH oxidase activation. J Clin Invest 1996; 97: 1916–1923.
24. Münzel T, Bassenge E. Long-term angiotensin-converting enzyme inhibition with high-dose enalapril retards nitrate tolerance in large epicardial arteries and prevents rebound coronary vasoconstriction in vivo. Circulation 1996; 93: 2052–2058.
25. Kurz S, Münzel T, Harrison DG. A role for angiotensin II in nitrate tolerance: chronic AT1 receptor blockade prevents development of tolerance and cross tolerance. Circulation 1995; 92 (suppl. I): 392.
26. Katz RJ, Levy WS, Buff L, et al. Prevention of nitrate tolerance with angiotensin-converting enzyme inhibitors. Circulation 1991; 83: 1271–1277.
27. Mehra A, Osterzega E, Shotan A, et al. Persistent hemodynamic improvement with short-term nitrate therapy in patients with chronic congestive heart failure already treated with captopril. Am J Cardiol 1992; 70: 1310–1314.
28. Pizulli L, Hagendorff A, Zirbes M, et al. Influence of captopril on nitroglycerin – mediated vasodilation and development of nitrate tolerance in arterial and venous circulation. Am Heart J 1996; 131: 342–349.
29. Muiesan ML, Boni E, Castellano M, et al. Effects of transdermal nitroglycerin in combination with an ACE inhibitor in patients with chronic stable angina pectoris. Eur Heart J 1993; 14: 1701–1708.
30. Dodak N, Makhone N, Flugelman MY, et al. Failure of captopril to prevent nitrate tolerance in congestive heart failure secondary to coronary artery disease. Am J Cardiol 1990; 66: 608–613.
31. Parker JD, Parker JO. Effect of therapy with an angiotensin converting enzyme inhibitor on hemodynamic and counterregulatory responses during continuous therapy with nitroglycerin. J Am Coll Cardiol 1993; 21: 1445–1453.
32. Elkayam U, Johnson JV, Shotan A, et al. Double-blind, placebo-controlled study to evaluate the effect of organic nitrates in patients with chronic heart failure treated with angiotensin-converting enzyme inhibition. Circulation 1999; 99: 2652–2657.
33. Cohn JN, Archibald DG, Ziesche S, et al. Effect of vasodilator therapy on mortality in chronic congestive heart failure. Results of a veterans administration cooperative study. New Engl J Med 1986; 314: 1547–1552.
34. Cohn JN, Johnson G, Ziesche S, et al. A comparison of enalapril with hydralazine-isosorbide dinitrate in the treatment of chronic congestive heart failure. New Engl J Med 1991; 325: 303–310.
35. Bauer JA, Fung HL. Concurrent hydralazine administration prevents nitroglycerine-induced hemodynamic tolerance in experimental heart failure. Circulation 1991; 84: 35–39.

36. Münzel T, Kurz S, Rajagopalan S, et al. Hydralazine prevents nitroglycerine tolerance by inhibiting activation of a membrane-bound NADH oxidase: a new action for an old drug. J Clin Invest 1996; 98: 1465–1470.
37. Holubarsch C. Klinische Ergebnisse der einzelnen therapeutischen Prinzipien: Vasodilatatoren. In: Autonomes Nervensystem und Herzinsuffizienz, (eds. R Griebenow, H Gülker, P Dominiak et al.) Georg Thieme Verlag Stuttgart-New York, 1995, 230–252.

Long-term oral treatment with nitrates – what is important?

M. Khan

Introduction

The clinical efficacy of nitrates has stood the test of time in the management of ischaemic heart disease. They have become the mainstay of immediate management of angina since the first use of Glyceryl Trinitrate (GTN) by William Murrell in 1879 (1).

First credit for the use of nitrate goes to Lauder Brunton, the House Physician at the Royal Infirmary in Edinburgh. He used amyl nitrite on a patient who complained of chest pain. Within 30 seconds and simultaneously with a flushing of the face, the pain completely disappeared (2). This report of almost instantaneous relief of anginal pain in his patient marked the introduction of one of the most effective classes of pharmacological agents in cardiovascular therapy.

Nitrates are often considered the "therapeutic fire extinguisher" and remain the favorite first line of treatment. The development of orally-administered, long-acting nitrates in the last two decades has been a major advance. They can provide not only relief of pain but also long-term prophylaxis against angina. Despite their use in the management of ischaemic heart disease for many years, we are still exploring new mechanisms of their action, and the basis of their action on healthy and diseased endothelium is now beginning to emerge. Their application in treating other forms of cardiovascular disease and the development of products with a more favorable pharmacological profile are being constantly sought.

Nitrates are rapidly absorbed from the gastrointestinal tract, mucus membrane and the skin. They are thus prescribed for oral, sublingual, buccal and transdermal use. There have been no controlled clinical trials to evaluate the long-term effects of oral nitrates, compared to placebos on morbidity.

The V-Heft trial for the treatment of heart failure, however, indicated symptomatic improvement and reduction in mortality of up to three years in moderately severe heart failure when nitrates were used along with hydralazine (3). The short-term benefit was control of symptoms and improvement of exercise tolerance, while the long-term benefit was moderate decrease in mortality. There was an improvement in symptoms and reduction in mortality of 38 % and 28 % over one and three years, respectively. The benefits of Isosorbide dinitrate and Hydralazine were greater in younger patients with low ejection fraction and in those with hypertension. In the second V–Heft trial, an Isosorbide dinitrate-Hydralazine combination was compared with Enalapril for moderate heart failure (4). The benefits of Enalapril over Isosorbide dinitrate-Hydralazine were evident over a five-year follow-up, but were mainly due to a decrease in the incidence of sudden death. There was, however, no difference between the two groups in mortality attributed to pump failure. The above data, although they do not provide conclusive evidence for the benefits of nitrate alone, do suggest some advantages of the drug when used in combination with Hydralazine in the management of congestive heart failure.

Oral nitrates are used mainly to provide symptomatic relief because of their anti-anginal and anti-ischaemic actions as manifested by the decrease in anginal frequency, improvement in exercise tolerance and prolongation of the walking distance prior to the development of angina. Despite the established use of these drugs, the clinical practice in many centers remains variable. In some, short-acting oral nitrates are the mainstays of anti-anginal therapy, while others with concerns for tolerance have a policy of using long-acting preparations with a nitrate-free period. Some are quite fond of using transdermal forms of nitrates.

Physiology of myocardial ischaemia

Ischaemia occurs when the oxygen supply to the myocardium is insufficient to meet the demand. Myocardial ischaemia results whenever the myocardium oxygen demand increases beyond the ability of coronary vasculature to deliver adequate blood supply or oxygen or when the supply decreases below myocardial oxygen demand as a result of coronary obstruction. This ischaemia is often associated with chest pain of angina pectoris or may be asymptomatic, as in silent ischaemia, but associated with ST-changes on the electrocardiogram.

It is also recognized that ischaemia may occur even if the obstruction is not very significant, since it depends to a great extent on the coronary vascular tone. If the tone of the coronary vasculature is increased, then the angina would occur even at minimal exertion. This was very well demonstrated by Maseri (5), indicating the influence of variation in the residual coronary artery flow reserve on the development of ischaemia even at a low level of activity.

Physiological effects of organic nitrates

Organic nitrates are believed to act directly on vascular smooth muscles through a complex biochemical pathway to produce vasodilatation. Nitrates are highly lipid soluble and are easily converted to nitric oxide (6). They enter the smooth muscle cells of the vessel wall, and once inside the cells, they are metabolized, releasing a nitrate intermediary that subsequently reacts with the sulfhydryl moiety to form an S-Nitrosothiol compound. It is this compound that – through liberation of the free radical nitric oxide - activates the enzyme Guanylate cyclase. This enzyme then converts the guanosine triphosphate (GTP) to cyclic guanosine monophosphate (cGMP), which lowers the cytoplasmic Ca+ by inhibiting calcium influx and finally resulting in vascular smooth muscle relaxation and vasodilatation (Figure 1).

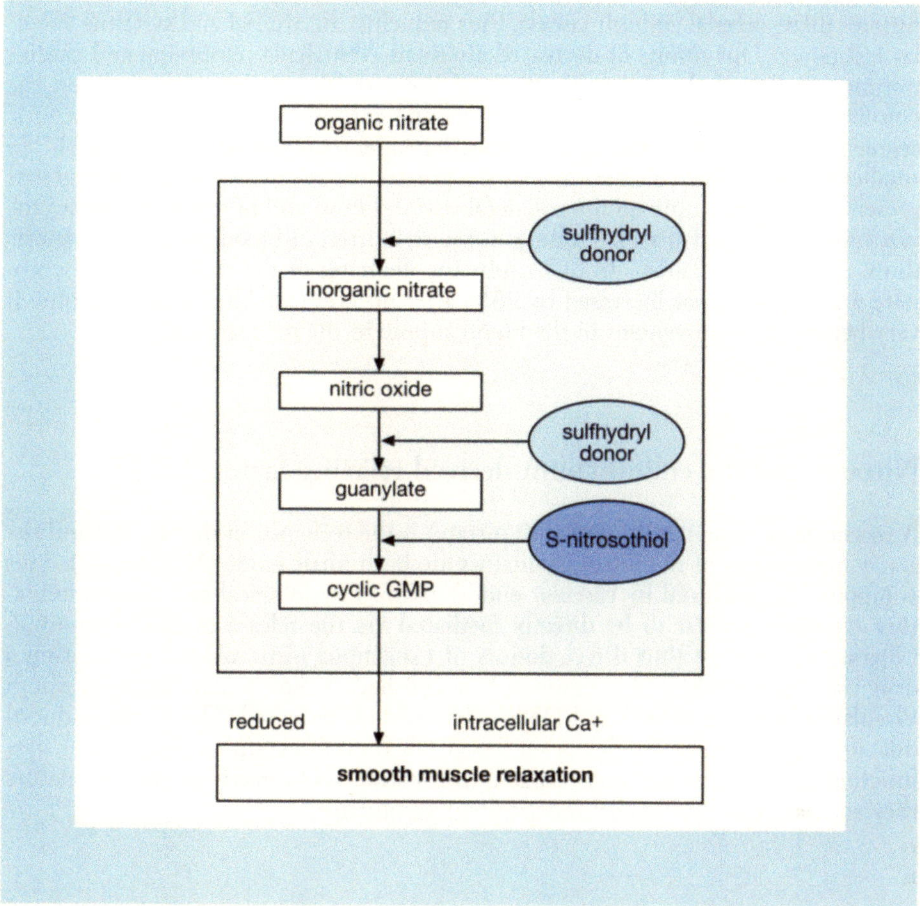

Figure 1: Metabolic pathway and mode of action of organic nitrates.

Haemodynamic effects of nitrates

Organic nitrates are potent relaxants of vascular smooth muscle and can affect capacitance, conductance and resistance vessels. The overall haemodynamic response to organic nitrate administration depends on several factors including dosage, vascular tone and left ventricular function. The main action of nitrates is venodilatation, resulting in reduced venous return, producing a marked fall in preload. As preload decreases, right and left ventricular end-diastolic pressures and volume decrease with reduction in myocardial work and oxygen requirement. The fall in ventricular diastolic pressure and wall tension in response to nitrates increases subendocardial blood flow. The net result is very favorable, with reduction in myocardial oxygen demand and an increase in subendocardial perfusion. At moderate to high dosages, nitrates dilate arterial conduit vessels, thus reducing the arterial and systemic vascular resistance. This results in decreased afterload. Ventricular emptying and cardiac output are improved. Coronary effects of the nitrates are exerted directly on the coronary vasculature by dilating the epicardial coronary arteries and the stenotic segment. The extent of the dilatation is dependent on the size of the vessel, the smaller the epicardial coronary artery, the greater the relative vasal dilator response present (7). It also improves the collateral size and flow, and prevents or reverses the vasoconstriction and spasm resulting in redistribution of blood flow to the epicardium and ischaemic areas. In one study, the diameter of the collateral-filled coronary artery vessels was increased by 38% after nitrates (8). The overall outcome is very beneficial improvement in the blood supply to the myocardium.

Nitrates and the endothelium-derived relaxing factor (EDRF)

A recent development with regard to nitrates is the recognition that EDRF and the active metabolites of the nitrovasodilators are both nitric oxide (NO) (9, 10). This compound is produced by vascular endothelium. Physiological and local vasodilatory responses appear to be directly mediated via the release of this compound. Organic nitrates are thus direct donors of exogenous nitric oxide, representing a drug class for treatment of patients with established cardiovascular disease in whom vascular response mediated by EDRF/nitric oxide is impaired. The nitrate-induced dilatation of coronary arteries occurs even when endothelium is denuded or dysfunctional (11). Through their capacity to provide NO directly to the vasculature they are classed as endothelial independent vasodilators.

Long-acting oral nitrate therapy

Isosorbide dinitrate (ISDN) and Isosorbide-5-mononitrate (ISMN) are the two most widely used compounds for oral therapy.

ISMN is the active metabolite of ISDN. Both products are available as immediate-release and slow-release preparations. Several studies have demonstrated the beneficial effects of the first dose of short acting ISDN on the Exercise Stress Test. However, with regular multiple doses of short-acting preparations, the development of tolerance is quite common. Conflicting reports have appeared in the literature. Danahy et al. (12) showed that an average dose of 29 mg (range 20-50 mg) of ISDN administration TID for six months continued to show beneficial effect on exercise tolerance. Similarly, Schneider et al. (13) found decreased angina frequency during long-term therapy with short-acting ISDN in a dose of 120 mg 4 times daily.

Parker et al. (14) also showed persistent efficacy and lack of tolerance with BID and TID dosage regimes for 30 mg of ISDN. Their treatment regime however, was asymmetrical, providing some nitrate-free intervals. Controlled-release ISDN formulations have been in use for the last two decades. Lee et al. (15) reported beneficial effects of controlled-release ISDN 60 mg QID, compared to placebo during long-term therapy. Tolerance was noted in another study reported by Silber et al. (16), but the tolerance was limited to an 80 mg dose and not when 160 mg was given once a day or BID.

It appears that a single daily dose of the sustained-release ISDN preparation is clearly effective and definitely a better way to administer than multiple dose regimens. The main active metabolite of ISDN is isosorbide-5-mononitrate (ISMN), which has a longer duration of action than its parent compound. ISMN has no active metabolite and it is nearly completely bioavailable after oral administration, since first pass hepatic metabolism is not involved (Table 1).

Table 1: Pharmacological characteristics of ISDN and ISMN (6).

Characteristics	ISDN	ISMN
bioavailability	20–40%	100%
hepatic first pass effect	necessary	none
G.I. absorption	rapid	slow
active metabolite	5-ISMN, 2-ISMN	none
half-life	1–2 hours	4–6 hours
duration of action	short	

Earlier studies showed the beneficial effects of 20–40 mg ISMN given on a BID, TID or QID basis. However, the development of tolerance was not uncommon (17–19). The development of tolerance to ISMN has been demonstrated to be less frequent when the drug was given in a dose of 20 mg in asymmetrical fashion (19–21). Another interesting feature of these studies of asymmetrically administered ISMN is that rebound nocturnal angina, or significant worsening in exercise duration prior to the morning dose, did not occur during long-term therapy (19–22). Long-acting ISMN preparations have been in use and they clearly provide additional benefits in terms of ease of therapy, longer duration and near total bioavailability of the product. Several studies of ISMN LA have shown not only immediate benefit in terms improvement in exercise capacity and reduction in the ST segment depression on exercise, but also in prophylaxis of angina and a continued effect for a period of up to one year (20–23).

A large-scale trial compared with placebo also demonstrated the benefits of 120–240 mg ISMN, but not the 30 or 60 mg sustained release formulation (24).

Nitrate tolerance

One limitation of long-term nitrate therapy is the development of nitrate tolerance. This phenomenon is not uncommon and the precise incidence of tolerance is not known. It was noted first in the exposure of munitions workers to nitrates. It is a major problem with frequent doses of short-acting preparations. The exact cellular mechanism of tolerance is complex and still unclear. Most evidence in patients with long-term angina and heart failure suggest that tolerance almost always occurs with those regimes that sustain high plasma and tissue levels of nitrates for more than 24 hours. Depletion of sulfhydryl groups, neuroendocrine activation, and an increase in the intravascular volume have all been implicated (Figure 2) (25).

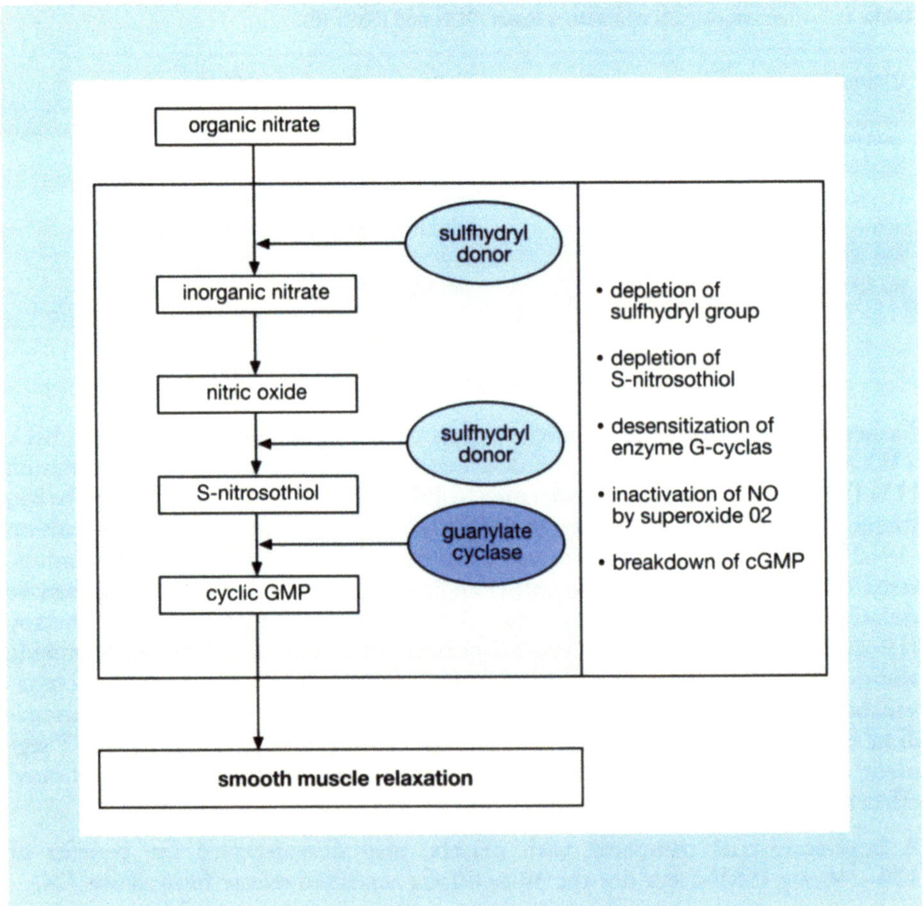

Figure 2: Proposed mechanisms for the development of nitrate tolerance.

Several potential therapeutic strategies to overcome this problem have been proposed (26), including administration of exogenous sulfhydryl compounds, blockade of neuroendocrine vasoconstriction, vasodilator therapy, i.e. hydralazine, and diuretics. A nitrate-free period of 8–10 hours during nitrate therapy remains most important, since continuously high nitrate concentrations in plasma and tissue over longer time periods remain the single most important cause of nitrate tolerance. A recent study suggests that the antioxidant effect of Carvedilol might prevent the development of nitrate tolerance (27). Another study found that nitrates enhanced the propensity for vasoconstriction, secondary to increased endothelin expression within the vascular smooth muscle, and noted that angiotensin A1 receptor blocker could abolish this phenomenon (27).

There are few adverse effects of long-acting nitrates. Headache is the major side effect during therapy, and most patients learn to live with mild headache. Severe headaches, requiring withdrawal from the drug, are uncommon and only occur in about 5–10% of patients. Nitrate-induced syncope has also been reported. Postural hypotension is also not uncommon during the first few days of therapy.

From our own experience, severe headache requiring withdrawal from nitrate therapy occurs in about 7% of patients.

Combination of long-acting nitrates

The most commonly used combination of long acting nitrates is a cardioselective long-acting beta-blocker. A combination of beta-blocker and nitrate would be appropriate, since they lower myocardial oxygen consumption and demand by different mechanisms. It is expected that the combination is likely to be superior to use of the single drugs. Several studies have documented beneficial effects of immediate release nitrate preparation when used along with beta-blockers, therefore, there is no reason why long acting preparations would not be useful in a similar situation (28). There are however, no controlled trials comparing the long-acting nitrate alone and in combination with a beta-blocker.

Recent studies in patients who remain symptomatic despite therapy with beta-blockers have shown substantial improvement during intermittent eccentric therapy with ISMN (28, 29). They demonstrated improvement in exercise duration and reduction in GTN consumption. Nitrates can also be combined with heart rate modulating calcium antagonist and their effects are likely to show additional benefits when compared to the either drug being used alone. The rationale for a combination therapy with nitrates in symptomatic or asymptomatic patients is that myocardial ischaemia will not be sufficiently controlled with monotherapy and may require more than one agent for significant reduction of their ischaemic episodes. In one study comparing nitrates, beta-blockers and calcium antagonists as monotherapy in patients with angina on exertion and at rest, no single drug was effective in all patients for elimination of anginal pain, ST-segment depression or other evidence of myocardial ischaemia (30). Combinations of anti-anginal therapy proved to be beneficial in such patients.

Combinational therapy has several advantages, foremost being the advantage of complimentary actions to potentially maximize the myocardial oxygen demand and minimize side effects. Nitrates exert their anti-ischaemic/anti-anginal actions through haemodynamic and direct action on coronary vasculature. The beta-blockers on the other hand, provide anti-ischaemic effect from direct depressant action on the heart, thus the combination will not only lower myocardial oxygen demand but will also improve myocardial oxygen supply. Combination therapy may counteract potentially adverse pharmacological effects of each agent when used alone. Nitrates, for example, will counteract the negative inotropic and chronotropic effects of beta-blockers.

The combination therapy may also be helpful in the avoidance of adverse effects of an individual drug by permitting the use of lower doses of each drug. For example, a patient with chronic obstructive airway disease, at risk of bronchial spasm when using beta-blockers, may be able to tolerate a lower dose of a cardioselective beta-blocker when used in combination with nitrates.

Coronary artery disease is a progressive disorder. It is therefore likely that a single drug/monotherapy may not remain effective over a longer period; the combination of anti-anginal agents with complementary actions will eventually be required to provide adequate relief or prevention of symptoms. Several studies have shown the beneficial effect of the addition of a second drug in the management of chronic stable angina (29–31). They were shown not only to benefit symptomatic episodes of angina, but in one study (32) the addition of a second or third drug resulted in substantial elimination of silent ischaemic episodes.

Antiplatelet effect

Platelet aggregation can cause vasoconstriction by stimulating release of Thromboxane A2 and other potent vasoconstrictors. There is increasing evidence that in addition to a direct relaxing effect on the vasculature smooth muscle, nitrates may also inhibit coronary vasoconstriction by inhibiting platelet aggregation. Nitrates can cause inhibition of platelet aggregation in response to injury to the arterial wall. Nitroglycerine inhibition of platelet aggregation has also been demonstrated in patients with coronary artery disease (33).

A recent study has shown that long-term therapy with oral nitrates does influence the natural history of coronary artery disease. Patients on long-term nitrates are less likely to have acute myocardial infarction than unstable angina on admission (34). Syndromes of acute myocardial ischaemia are often associated with focal or generalized endothelial dysfunction (35, 36). Platelets often initiate a thrombotic cascade in such patients. Nitrate action will lead to decreased coronary artery spasm or vasoconstriction and platelet inhibition, particularly at or near the site of internal disruption. Clearly this anti-platelet effect, along with important decrease in myocardial oxygen consumption and increase in the coronary blood flow, would be significantly salutary.

Although it is difficult to attribute the beneficial contribution of nitrates in chronic stable angina to anti-platelet effects, there is no doubt that a significant anti-platelet and anti-thrombotic effect is of importance in patients with unstable coronary syndromes. Endothelial dysfunction is commonly associated with these syndromes and platelets are involved in the initiation of the thrombotic cascade, leading to intraluminal thrombus formation and platelet-thrombin activation.

Chronobiology of ischaemic events

Circadian variations should be taken into account in the management of coronary artery disease. It has become clear that pathophysiological events within the cardiovascular system do not occur at random. These events are not evenly distributed. Ambulatory ST-segment monitoring has facilitated the study of observing the ST-segment changes over 24 hours.

Anginal symptoms of disturbed coronary artery circulation are most frequently reported during the morning after arising. There is strong evidence that long-acting nitrates can attenuate this circadian variation of anginal attacks. In patients with chronic stable angina, the circadian pattern demonstrates a peak during the morning hours, similar to the pattern observed in myocardial infarction. This peak is much more evident in patients when linked to the specific time of awakening and commencing activities. A recent European survey on circadian variation of angina pectoris (ESCVA) (37) demonstrated that the maximum number of anginal attacks occurred within two hours after awakening. Approximately 50% of all angina attacks occurred within 6–8 hours after waking. There was very little angina during the nighttime. A drug likely to be effective for 12–14 hours, with a peak of 3–4 hours after intake, is likely to be singularly beneficial in the management of such patients.

Elantan LA, a long-acting Isosorbide-5-mononitrate by virtue of its unique release mechanism, provides the desired pharmacokinetics. It allows the peak and sustained levels to be reached at a time when anginal episodes are most likely to occur.

Awareness of this chronobiology is therefore very important for the successful management of stable angina pectoris.

Conclusion

We have come a long way since the early use of nitrates in 1879. Multiple beneficial effects of nitrates are now well recognized. Our understanding and knowledge of the phenomenon of tolerance is improving and strategies to eliminate or attenuate nitrate tolerance are constantly sought. The ability to overcome this problem would represent a major advance in the therapy of angina and congestive heart failure. We are also aware of the individual variations of response to therapy and the fact that in some patients relief of angina may persist despite development of haemodynamic tolerance. Nitrates, by virtue of their multiple actions, are considered to be the first line of therapy, either alone or in combination. Our long experience suggests that the therapy could be initiated in the majority of patients because of its reassuring efficacy, multiple modes of action and proven long-term safety record. The choice of therapeutic agents should be made on an individual basis. With skillful galenics, therapy can be tailored so as to provide maximum benefit not only for exertional angina but also for episodes of silent ischaemia related to variable levels of activity. Furthermore, long-term nitrate administration in patients with coronary artery disease and congestive heart failure may also be beneficial by favorably altering the natural history of the underlying disease process.

References

1. Murrell W: Nitroglycerine as a remedy for angina pectoris. Lancet Jan 18, 1879; 81–81, 113–115, 151–152.
2. Brunton TL: On the use of nitrite of amyl in angina pectoris. Lancet July 27, 1867; 97–98.
3. Cohn JN, Archibald DG, Ziesche S, et al. Effect of vasodilator therapy on mortality in chronic congestive heart failure. Results of a veterans administration cooperative study. New Engl J Med 1986; 314: 1547–1552.
4. Cohn JN, Johnson G, Ziesche S, et al. A comparison of enalapril with hydralazine-isosorbide dinitrate in the treatment of chronic congestive heart failure. New Engl J Med 1991; 325: 303–310.
5. Maseri A, Chierchia S, Kaski JC. Mixed angina pectoris. Am J Cardiol 1985; 56: 30E–33E.
6. Murad F. Drugs used for the treatment of angina: organic nitrates, calcium channel-blockers and B-adrenergic antagonists. In: Goodman and Gilman. The Pharmacological Basis of Therapeutics (eds. AG Gilman, TW Rall, AS Nies et al.), Pergamon Press, New York, 1990, 764–783.
7. Brown BG. Response of normal and diseased epicardial coronary arteries to vasoactive drugs: quantitative arteriographic studies. Am J Cardiol 1985; 56: 23E–29E.
8. Brown BG, Bolson E, Peterson RB, et al. The mechanisms of nitroglycerin action: stenosis vasodilatation as a major component of the drug response. Circulation 1981; 68: 1089–1097.
9. Ignarro LJ, Buga GM, Wood KS, et al. Endothelium derived relaxing factor produced and released from artery and vein is nitric oxide. Proc Natl Acad Sci U.S.A. 1987; 84: 9265–9269.
10. Khan MT, Furchgott RF. Similarities of behavior of nitric oxide (NO) and endothelium-derived relaxing factor in a perfusion cascade bioassay system. Fed Proc 1987; 46: 385 (abstract).
11. Lüscher TF. Endothelial-derived nitric oxide: the endogenous nitrovasodilator in the human cardiovascular system. Eur Heart J 1991; 12 (suppl. E): 2–11.
12. Danahy DT, Aronow WS. Haemodynamic and antianginal effects of high dose oral isosorbide dinitrate after chronic use. Circulation 1977; 56: 205–212.
13. Schneider WU, Bussman WD, Stahl B, et al. Dose-response relation of antianginal activity of isosorbide dinitrate. Am J Cardiol 1984; 53: 700–705.
14. Parker JO, Farrell B, Lahey KA, et al. Effect of intervals between doses on the development of tolerance to isosorbide dinitrate. N Engl J Med 1987; 316: 1440–1444.

15. Lee G, Mason DT, Amsterdam EA, et al. Antianginal efficacy of oral therapy with isosorbide dinitrate capsules: prolonged benefits shown by exercise testing in patients with ischaemic heart disease. Chest 1978; 73: 327–332.
16. Silber S, Vogel AC, Krause KH, et al. Induction and circumvention of nitrate tolerance applying different dosage intervals. Am J Med 1987; 83: 860–870.
17. Kohli RS, Rodrigues EA, Kardash, MM, et al. Acute and sustained effects of isosorbide-5-mononitrate in stable angina pectoris. Am J Cardiol 1986; 58: 727–731.
18. Seabra-Gomes R, Aleixo AM, Adao M, et al. Comparison of the effects of a controlled-release formulation of isosorbide-5-mononitrate and conventional isosorbide dinitrate on exercise performance in man with stable angina pectoris. Am J Cardiol 1990; 65: 1308–1312.
19. Thadani U, Maranda C, Amsterdam E, et al. Lack of pharmacological tolerance and rebound angina pectoris during twice-daily therapy with isosorbide-5-mononitrate. Ann Intern Med 1994; 120: 353–359.
20. Ahmadinejad M, Eghbal B, Sorgenicht W, et al. Slow-release isosorbide-5-mononitrate – a new once-daily therapeutic modality for angina pectoris. Eur Heart J 1988; 9 (suppl. A): 135–139.
21. Parker JO. Nitrates and angina pectoris. Am J Cardiol 1993; 72: 3C–8C.
22. Parker JO. Eccentric dosing with isosorbide-5-mononitrate in angina pectoris. Am J Cardiol 1993; 72: 871–876.
23. Wisenberg G, Roks C, Nichol P, et al. Sustained effect of and lack of devolpment of tolerance to controlled-release isosorbide-5-mononitrate in chronic stable angina pectoris; Am J Cardiol 1989; 64: 569–576.
24. Chrysant SG, Glasser SP, Bittar N, et al. Efficacy and safety of extended release isosorbide mononitrate for stable effort angina pectoris. Am J Cardiol 1993; 72: 1249–1256.
25. Münzel T, Kurz S, Heitzer T, et al. New insights into mechanisms underlying nitrate tolerance. Am J Cardiol 1996; 77: 24C–30C.
26. Mangione NJ, Glasser SP. Phenomenon of nitrate tolerance. Am Heart J 1994; 128: 137–146.
27. Glasser SP. Prospects for therapy of nitrate tolerance. Lancet 1999; 353: 1545–1546.
28. Cohn PF. Concomitant use of nitrates, calcium channel blockers and beta-blockers for optimal antianginal therapy. Clin Cardiol 1994; 17: 415–421.
29. Silber S, Vogler AC, Spiegelsberger F, et al. The insufficient nitrate response: Patients' characterization and response to beta and calcium blockade. Eur Heart J 1988; 9 (suppl. A): 125–134.
30. Quyyumi AA, Crake T, Wright CM, et al. Medical treatment of patients with severe exertional and rest angina: double blind comparison of beta-blocker, calcium antagonist and nitrate. Br Heart J 1987; 57: 505–511.
31. Akhras F, Jackson G. Efficacy of nifedipine and isosorbide mononitrate in combination with atenolol in stable angina. Lancet 1991; 338: 1036–1039.
32. Koehn DK, Glasser SP. Impact of antianginal drug therapy on asymptomatic myocardial ischaemia. J Clin Pharmacol 1989; 29: 722–727.
33. Diodati J, Théroux P, Latour JG, et al. Effects of nitroglycerine at therapeutic doses on platelet aggregation in unstable angina pectoris and acute myocardial infarction. Am J Cardiol 1990; 66: 683–688.
34. Garcia-Dorado D, Permanyer-Miralda G, Brotons C, et al. Attenuated severety of new acute ischaemic events in patients with previous coronary heart disease receiving long-acting nitrates. Clin Cardiol 1999; 22: 303–308.
35. Abrams J. Beneficial actions of nitrates in cardiovascular disease. Am J Cardiol 1996; 77: 31C–37C.
36. Lam JYT, Chesebro JH, Fuster V. Platelets, vasoconstriction and nitroglycerine during arterial wall injury: a new antithrombotic role for an old drug. Circulation 1988; 78: 712–716.
37. Willich SN. European Survey on Circadian Variation of Angina Pectoris (ESCVA) – wake-time adjusted morning risk. Eur Heart J 1999; 20 (suppl.): 164.

Alternatives in long-term nitrate therapy

S. Ghio, C. Raineri

Introduction

Organic nitrates are the oldest of all the anti-anginal drugs, having been used for more then 100 years to treat myocardial ischaemia. The mechanism of action of nitrates is unique in that they act on both the demand and on the supply side of the myocardial oxygen consumption balance by reducing left ventricular preload and afterload, by dilating epicardial coronary arteries and by improving flow through collateral vessels (1). These drugs are available in a variety of formulations with different routes of administration. When we talk of long-term therapy with nitrates, whichever preparation is used, a correct dosing strategy is the key to avoid the loss of the acute effects that occurs during continuous administration.

Oral nitrates

The most widely used long-acting oral nitrates contain either isosorbide dinitrate (ISDN) or isosorbide mononitrate (ISMN), which is in fact the main active metabolite of ISDN. ISMN has a longer duration of action than ISDN and is nearly 100% bioavailable after oral administration.

When immediate-release ISDN is given four times daily at evenly spaced intervals, tolerance quickly develops; when ISDN is taken on an eccentric schedule twice daily or three times daily, tolerance does not develop to the first dose, but the effects of the second and of the third doses are reduced (2–4). The development of tolerance during long-term treatment with extended-release ISDN given once or twice daily has not been adequately evaluated.

ISMN products are also available in both immediate-release and extended-release formulations. Immediate-release ISMN can be administered according to an eccentric dosing regimen to avoid the development of tolerance: when ISMN is given b.i.d., 7 hours apart, with a 17-hour dose-free interval after the second daily dose, exercise duration is prolonged after both the first and second dose, although a slight attenuation of effect occurs over time (5, 6).

Although several studies suggested maintained improvement in exercise duration for as much as 12 hours using 60 mg preparations, in a larger trial only the higher doses of 120 and 240 mg showed significant long-term effects on total exercise duration (7–9).

Transdermal nitrates

Long-term therapy with nitroglycerin (NTG) patches is an important alternative to oral nitrates: in fact, compliance to therapy is generally improved when drugs are administered transdermally (10). However, despite their worldwide popularity, the efficacy of NTG patches has been questioned in a two-front attack: continuous treatment cannot be recommended because of complete tolerance development, and intermittent therapy is problematic because of rebound phenomena.

Tolerance during continuous treatment with NTG patches

A number of randomized trials with exercise tests have addressed this question. The results of the Transdermal Nitroglycerin Cooperative Study indicate that tolerance develops in most patients within 24 hours and that the loss of the effects cannot be overcome by larger doses (11). In several smaller studies, intermittent and continuous treatment have been challenged: usually, the results obtained 2–5 hours after repeated patch application indicate a greater effect on total exercise duration for intermittent treatment (12–17). Hence, if nitroglycerin patches are given with the primary intention of improving exercise capacity, an intermittent treatment regimen should be recommended. However, despite what is commonly acknowledged, even continuous treatment may result in positive effects in some patients.

Three double-blind studies reported maintained efficacy (partial tolerance) of continuous treatment with NTG patches (13, 18, 19). These are not necessarily false-positive results: the differences with the Cooperative Study can be explained by differences in the study population. When patients show a greater initial response to the NTG treatment and a small placebo effect is noticed during the trial, there is obviously a much better chance to demonstrate a moderate effect during sustained NTG therapy.

Finally, although the most accepted way to assess treatment efficacy in patients with stable angina pectoris is through exercise tests, one must consider that daily-life ischaemic episodes are due to a combination of two physiopathologic mechanisms: increased oxygen demand and reduced oxygen supply. This means that, when we evaluate patients with stable angina, the frequency of spontaneous anginal attacks and the total ischaemic burden at Holter monitoring are relevant parameters in addition to those derived by the exercise test and we should not be surprised if such measurements even suggest different conclusions (20). It is therefore of interest that in a large open trial involving more than 2,400 patients, continuous NTG administration was found to be as effective as oral nitrates in reducing anginal attacks during daily life (21). We can conclude that, if the objective of medical therapy is to improve quality of life of stable angina patients at low risk for cardiac events, continuous treatment with NTG patches is one of the options to be considered.

In summary, the evidence pointing toward a complete tolerance development during continuous NTG treatment is not so impressive: long-term responsiveness can be observed in those patients who show an initial greater effect and it is possible that in each patient tolerance occurs to some, but not all, drug effects (for example the acute effects on exercise duration may disappear while the patient still experiences a reduction of anginal episodes during daily life). The concept that nitrate tolerance is not an "all or none" phenomenon is more than a simple hypothesis: it is supported by hemodynamic studies that have shown a differential development of nitrate tolerance in the venous and arterial side of the circulation and a dissociation between the loss of the clinical (ergometric) effects and the persistence of the arterial hemodynamic action of the drug (22, 23, 24).

Problems with intermittent therapy?

Although an increased frequency of episodes of rest angina during patch-off hours has been reported and might be a cause for concern, no such effects were reported in other studies, including a larger trial of intermittent NTG therapy (14, 25–28). Therefore, no firm conclusion can be drawn on the risk of rebound phenomena during nitrate free periods on the basis of the available data.

The so-called "zero-hour" effect, which is a reduction in exercise capacity before the daily patch application compared to placebo, is probably not so clinically relevant if the patch-off hours are limited to the night-time period (29).

Conclusions

Nitrate tolerance attenuates the clinical effects of all known nitrate regimens some hours after administration, but the magnitude of the persistent efficacy may vary between patient populations and according to the method used for efficacy evaluation. For the majority of stable angina patients on transdermal nitrates, intermittent therapy is preferable, since it is superior in improving exercise capacity. However, patients with a high nitrate responsiveness and/or frequent spontaneous anginal attacks may benefit more from continuous therapy. Rebound phenomena have been described during intermittent therapy but are clinically relevant only in a minority of the patients.

The present evidence suggests that efficacy and tolerance are comparable for NTG patch therapy and for oral nitrates when these drugs are administered according to current principles.

References

1. Parker JD, Parker JO. Nitrate therapy for stable angina pectoris. N Engl J Med 1998; 338: 520–531.
2. Thadani U, Fung H-L, Darke AG, et al. Oral isosorbide dinitrate in angina pectoris; comparison of duration of action and dose-response relation during acute and sustained therapy. Am J Cardiol 1982; 49: 411–419.
3. Parker JO, Farrel B, Lahley KA, et al. Effect of interval between doses on the development of tolerance to isosorbide dinitrate. N Engl J Med 1987; 316: 1440–1444.
4. Bassan MM. The daylong pattern of the antianginal effect of long-term three times daily administered isosorbide dinitrate. J Am Coll Cardiol 1990; 16: 936–940.
5. Parker JO and the Isosorbide-5-Mononitrate Study Group. Eccentric dosing with isosorbide-5-mononitrate in angina pectoris. Am J Cardiol 1993; 72. 871–876.
6. Thadani U, Maranda CR, Amsterdam E, et al. Lack of pharmacologic tolerance and rebound angina pectoris during twice-daily therapy with isosorbide-5-mononitrate, Ann Intern Med 1994; 120: 353–359.
7. Parker JO. Controlled release isosorbide-5-monitrate in angina pectoris: A comparison with standard formulation isosorbide dinitrate. Can J Cardiol 1991; 7: 125–130.
8. Wisenberg G, Roks C, Nichol P, et al. Sustained effect of and lack of devolpment of tolerance to controlled-release isosorbide-5-mononitrate in chronic stable angina pectoris; Am J Cardiol 1989; 64: 569–576.
9. Chrysant SG, Glasser SP, Bittar N, et al. Efficacy and safety of extended release isosorbide mononitrate for stable effort angina pectoris. Am J Cardiol 1993; 72: 1249–1256.
10. Nissinen A, Wiklund I, Lahti T, et al. Anti-anginal therapy and quality of life. A comparison of the effects of transdermal nitroglycerin and long-acting oral nitrates. J Clin Epidemiol 1991; 44: 989–997.
11. Steering Committee, Transdermal Nitroglycerin Cooperative Study. Acute and chronic efficacy of continuous twenty-four hour application of transdermal nitroglycerin. Am J Cardiol 1991; 68: 1263–1273.
12. Luke R, Sharpe N, Coxon R. Transdermal nitroglycerin in angina pectoris: Efficacy of intermittent application. J Am Coll Cardiol 1987; 10: 642–646.
13. Milliano PA, Koster RW, Bar FW, et al. Long-term efficacy of continuous and intermittent use of transdermal nitroglycerin in stable angina pectoris, Am J Cardiol 1991; 68: 857–862.
14. Ferratini M, Pirelli S, Merlini P, et al. Intermittent transdermal nitroglycerin monotherapy in stable exercise-induced angina: A comparison with a continuous schedule, Eur Heart J 1989; 10: 998–1002.
15. Rossetti E, Luca C, Bonetti F, et al. Transdermal nitroglycerin reduces the frequency of anginal attacks but fails to prevent silent ischemia. J Am Coll Cardiol 1993; 21: 337–342.
16. Gumbrielle T, Freedman SB, Fogarty L, et al. Efficacy, safety, and duration of nitrate-free interval to prevent tolerance to transdermal nitroglycerin in effort angina. Eur Heart J 1992; 13: 761–768.
17. Cowan J, Bourke J, Reid D, et al. Prevention of tolerance to nitroglycerin patches by overnight removal. Am J Cardiol 1987; 60: 271–275.
18. Greco R, Schiattarella M, Wolff S, et al. Long-term efficacy of nitroglycerin patch in stable angina pectoris. Am J Cardiol 1990; 65: 9J–15J.
19. Scardi S, Camerini F, Pandullo G, et al. Efficacy of continuous and intermittent transdermal treatment with nitroglycerin in effort angina pectoris: A multicentric study. Int J Cardiol 1991; 32: 241–248.
20. Shell WE, Dobson D. Dissociation of exercise tolerance and total myocardial ischemic burden in chronic stable angina pectoris. Am J Cardiol 1990; 66: 42–48.

21. Savonitto S, Motolese M, Agabiti-Rosei E. Antianginal effect of transdermal nitroglycerin and oral nitrates given for 24 hours a day in 2,456 patients with stable angina pectoris. The Italian Multicenter Study, Int J Clin Pharmacol Ther 1995; 33: 194–203.
22. Stewart DJ, Elsner D, Sommer O, et al. Altered spectrum of nitroglycerin action in long-term treatment: nitroglycerin-specific venous tolerance with maintenance of arterial vasodepressor potency. Circulation 1986; 74: 573–582.
23. Ghio S, De Servi S, Perotti R, et al. Different susceptibility to the development of nitroglycerin tolerance in the arterial and venous circulation in humans. Effects of N-acetylcysteine administration, Circulation 1992; 86: 798–802.
24. Klemsdal TO, Mundal HH, Gjesdal K. Mechanisms of tolerance during treatment with nitroglygerin patches for 24 h. Eur J Clin Pharmacol 1996; 51: 227–230.
25. Pepine CJ, Lopez LM, Bell DM, et al. Effects of intermittent transdermal nitroglycerin on occurence of ischemia after patch removal: Results of the Second Transdermal Intermittent Dosing Evaluation Study (TIDES-II). J Am Coll Cardiol 1997; 30: 955–961.
26. Fox KM, Dargie HJ, Deanfield J, et al. Avoidance of tolerance and lack of rebound with intermittent dose titrated transdermal glyceryl trinitrate, Br Heart J 1991; 61: 151–155.
27. Holdright DR, Katz RJ, Wright CA, et al. Lack of rebound during intermittent trandermal treatment with glyceryl trinitrate in patients with stable angina on background betablocker. Br Heart J 1993; 69: 223–227.
28. Parker JO, Amies MH, Hawkinson RW, et al. Intermittent transdermal nitroglycerin therapy in angina pectoris. Clinically effective without tolerance or rebound, Circulation 1995; 91: 1368–1374.
29. Demots H, Glasser SP. Intermittent transdermal nitroglycerin therapy in the treatment of chronic stable angina. J Am Coll Cardiol 1989; 13: 786–793.

The Yin and Yang of nitric oxide, or: the importance of vascular control in acute ischaemia and decrease in cardiac output

G. Cotter

Introduction

Nitric oxide (NO) is one of the most important endogenous regulators of vascular tone. Nitrates, which enhance the activity of this pathway, have been used for more than 100 years in the treatment of ischaemia and heart failure. Despite their long use the exact mechanism of action of nitrates in acute cardiac conditions has not been fully elucidated.

We recently performed a series of studies based on the assumption that during acute ischaemia and decrease in cardiac output activation of vascular mediators (including NO) regulates the vascular response to these acute cardiac conditions determining whether the patient's condition will stabilise or deteriorate. The nitrate's main effect is enhancing the NO pathway, reducing systemic vascular resistance and afterload, reducing preload, improving myocardial relaxation, re-establishing balance between cardiac demand and coronary supply, and possibly aborting ischaemia.

Hypothesis

In patients who suffer from acute ischaemia, especially if accompanied by decreased cardiac output, few mediators are activated that have a primarily (although not exclusive) vascular modulating effect. These mediators are of two important classes. The first: vasoconstrictors including cathecholamines, angiotensin II and endothelin. The second includes peptides that are primarily vasodilators and diuretics, including atrial natriuretic peptide, anti-diuretic hormone and nitric oxide. The role of all these mediators includes:

1. Aborting ischaemia.
2. Re-establishing the balance between myocardial demand and supply.
3. Preventing the steep increase in left ventricular end diastolic pressure (LVEDP) induced by ischaemia, which can lead to pulmonary oedema, and reduced oxygenation.
4. Redistribution of the reduced cardiac output from non-vital to vital organs (e.g. heart, brain and kidneys).

This role is achieved by four main mechanisms:

1. Ischaemia abortion: in ischaemic regions due to the fixed significant obstruction in an epicardial coronary artery the coronary flow to the ischaemic region is jeopardised. Therefore, two mechanisms are activated:
   I) Arteriolodilatation in the ischaemic region, which serves to divert blood flow to the ischaemic myocardium.
   II) Increased aortic root blood pressure increasing coronary pressure directly, and therefore coronary flow.
2. Re-establishing the balance between myocardial demand and supply: this is achieved by local release of myocardial depressant factors in the ischaemic region, which decreases myocardial systolic function thereby inducing stunning. This serves to protect the ischaemic myocardium from further ischaemic damage.
3. Reducing LVEDP: the release of some mediators directly improves relaxation, therefore reducing LVEDP and preventing pulmonary oedema. This prevents the reduced oxygenation and increased breathing work that occur during pulmonary oedema, which further decreases cardiac output and increases demand.
4. Redistribution of cardiac output to vital organs: this is achieved by selective and balanced vasoconstriction of resistance arteries in non-vital organs including the skin, muscles and splanchnic bed. On the other hand, the blood flow to the brain, heart and kidneys is not reduced but rather increased. This effect is achieved through the balanced action of all mediators.

Nitric oxide (NO) has an important role in the delicate balance between all known (and unknown) mediators. In acute ischaemia it is released both locally and systemically (especially if cardiac output is also decreased). Locally it induces coronary arteriolodilatation therefore improving coronary flow. It has a significant cardiodepressant effect, therefore reducing the myocardial systolic work in ischaemic regions and preventing further ischaemic damage to the myocardium. Furthermore, it improves myocardial relaxation therefore reducing LVEDP and preventing pulmonary oedema. The prevention of pulmonary oedema is further achieved by the systemic effects of NO, including venodilatation (and reduction of pre-load) and arteriodilation, which prevents excessive increase in systemic vascular resistance in response to the decrease in cardiac output and prevents excessive increase in afterload.

However, contrary to common wisdom, the release of these compensatory neurohormones, although usually adequate, leading to stabilisation of both ischaemia and decreased cardiac output, might be excessive or suboptimal in some cases. This may be the reason why in some patients, despite similar amounts of ischaemia and decreased cardiac performance, the clinical course is complicated by pulmonary oedema, hypotension and cardiogenic shock while in others the same myocardial performance is associated with stabilisation and recovery.

The main obstacle to the study of these issues is the significantly different response of these mediators to ischaemia and decreased cardiac output in different animal models. Therefore, we have studied the effect of manipulation of the NO pathway in three different acute cardiac scenarios: acute ischaemia, pulmonary oedema and cardiogenic shock.

Acute ischaemia

Patients and methods

Included in this prospective study were 72 patients who complained of chest pain and were transported to the emergency room by a mobile intensive care unit because of unstable angina (1). Inclusion criteria were the presence of typical chest pain suggestive of myocardial ischaemia lasting for longer than 10 minutes, and ST-depressions of 1 mm or more, compatible with myocardial ischaemia, on two contiguous ECG leads. Exclusion criteria were baseline systolic blood pressure equal to or less than 100 mmHg, ST-elevation on ECG, left ventricular hypertrophy on ECG, clinical or electrocardiographic evidence of right ventricular ischaemia.

Patients were alternately assigned to receive ISDN according to one of the following protocols:

1. Group A (n=38) intravenous bolus of 1 mg (1,000 microgram, 1 cc) ISDN repeated every 3 minutes.
2. Group B (n=34) sublingual ISDN administered as a 5 mg spray every 10 minutes.

Treatment was titrated to achieve symptom alleviation or reduction of blood pressure by up to 20%.

A detailed questionnaire reflecting the patient's condition was completed by the physician and paramedic in the mobile intensive care unit. The recorded data included grading of chest pain on a scale of 0 to 3 (0 = no pain, 1 = mild pain, 2 = moderate pain, and 3 = severe pain). ECG was performed upon recruitment and at 1 hour in the emergency room.

Results and discussion

Although i.v.-treated patients required less ISDN than SL-treated patients (7.0 + 3.4 mg vs. 8.7 + 3.8, $p = 0.06$), the average blood pressure reduction was similar in the two groups and ischaemia reduction was substantially more pronounced in the i.v.-treated patients (Table 1). This could be explained by the significant relationship between an exact decrease in blood pressure and ischaemia alleviation. As depicted in Table 1, the optimal ischaemia reduction was achieved in patients in whom blood pressure was decreased by 5–20%. In patients in which the blood pressure reduction was sub-optimal (<5%), and in patients in whom blood pressure reduction was excessive (>20%), ischaemia control was reduced. The exact titration of ISDN dose against blood pressure reduction was of course best achieved by small repeated i.v. boluses (Table 1), therefore explaining the superiority of this treatment as compared to SL administration of ISDN. The different effect of a fixed SL dose of ISDN in different patients is probably related to the highly variable bioavailability of ISDN delivered as SL spray (58±29%) (2), inducing unpredictable effects on blood pressure and therefore sub-optimal ischaemia control.

Table 1: Haemodynamic and clinical parameters in patients with acute ischaemia during treatment with nitrates (1).

	i.v.-treated group			SL-treated group		
	before treatment	after treatment	change during treatment	before treatment	after treatment	change during treatment
systolic blood pressure (mmHg)	160 ± 24	134 ± 18	26 ± 15 (16%)	159 ± 23	137 ± 38	22 ± 16 (14%)
diastolic blood pressure (mmHg)	94 ± 14	81 ± 11 (14%)	13 ± 10	93 ± 14	81 ± 16 (13%)	12 ± 8
pain score (on scale of 0-3)	2.2 ± 0.6	0.7 ± 0.5	-1.5 ± 0.8 (67%)	2.2 ± 0.8	1.3 ± 0.9	-0.9 ± 0.9 (39%)
R wave summation (mm)	80 ± 30	85 ± 26	5 ± 16	77 ± 25	75 ± 28	-2 ± 16 (3%)
ST-segment summation (mm)	9 ± 6.6	3.9 ± 4.1	-5.1 ± 3.6 (58%)	7.9 ± 3.7	5.8 ± 3.6	-2.2 ± 2.7 (27%)

The results of this study suggest that the vascular control is important in the rapid abortion of acute ischaemia. However, this response is not fixed but rather varies from patient to patient. Therefore, the "optimal" nitrate dose is also individual. A "fixed recipe" for nitrate treatment does not exist. Accordingly, an optimal response is achieved through individual titration of ISDN dose against an optimal blood pressure reduction. This is best achieved by administration of repeated boluses of i.v. ISDN 1 mg under strict blood pressure monitoring.

Pulmonary oedema

In continuation to our hypothesis (3), we postulated that pulmonary oedema is caused by an imbalance between the different mediators released during an acute decrease in cardiac performance (Figure 1). This imbalance favours an excessive peripheral vasoconstriction, which leads to a steep increase in afterload, further decreasing cardiac output. Furthermore, this imbalance leads to a suboptimal effect on cardiac relaxation that, combined with significantly decreased cardiac output, leads to a sustained increase in LVEDP and pulmonary oedema. Therefore, this maladaptive mechanism could be affected by increasing the effect of NO by administration of i.v. high-dose ISDN or administration of endothelin antagonists. We have tested the first possibility by comparing the administration of high-dose i.v. ISDN to conventional treatment in patients with severe pulmonary oedema.

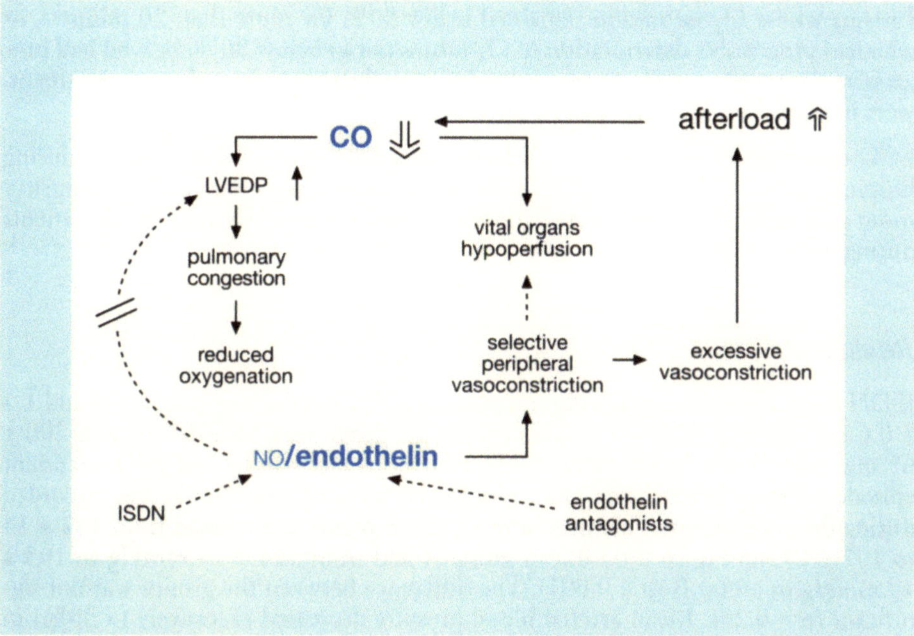

Figure 1: Pulmonary oedema: an excessive vasoconstrictive reaction causes an increase in afterload. The resulting decrease in cardiac output (CO) leads finally to pulmonary oedema.

Patients and methods

Included in this prospective study were 100 patients with symptoms and signs of pulmonary oedema that was confirmed by chest x-ray findings in the emergency room, and O_2-saturation of less than 90%, measured by pulse oximetry prior to O_2-administration, with the patient in the sitting position. Exclusion criteria were blood pressure of less than 110/70 mmHg and previous adverse reaction to furosemide or nitrates.

On admission, each patient was placed in the sitting position and oxygen was administered by facemask with a rebreathing bag at a rate of 10 l/min. i.v. boluses of morphine 3 mg and furosemide 40 mg were administered. Patients were randomised to receive:

1. Group A (n = 52): i.v. 3 mg bolus of ISDN every 5 minutes.
2. Group B (n = 52): i.v. 80 mg bolus of furosemide every 15 minutes and ISDN 1 mg/h (16 µg/min) increased by 1 mg/h every 10 minutes.

Treatment was continued in both groups until O_2-saturation increased to 96% or mean arterial blood pressure decreased by 30% or to lower than 90 mmHg.

Patients whose O_2-saturation remained below 80% for more than 20 minutes, or who had progressive deterioration of O_2-saturation to below 80%, or who had progressive dyspnoea, apnea, or severe brady- or tachyarrhythmias despite treatment, were intubated and mechanically ventilated.

ECG examination was repeated 24 hours after admission and as required during hospitalisation. CPK values were determined upon admission to the emergency room and again 24 hours later. Echocardiography was performed on all patients during hospitalisation.

Results and discussion

ISDN doses administered during treatment were 11.4 ± 6.8 mg in group A and 1.4 ± 0.6 mg in group B ($p < 0.001$). Furosemide doses were 56 ± 28 mg and 200 ± 65 mg, respectively ($p < 0.001$). Aside from sinus tachycardia and mild, transient episodes of sinus bradycardia, no severe brady- or tachyarrhythmias were recorded during drug treatment. The mean arterial blood pressure decreased from 132 ± 14 to 107 ± 15 mmHg ($p < 0.001$) in group A and from 124 ± 24 mmHg to 103 ± 19 mmHg in group B ($p < 0.001$). The difference between the groups was not significant ($p = 0.26$). Mean arterial blood pressure decreased excessively (> 30%) in five patients in group A (10%) and in seven (13%) in group B, but none of these patients showed a decrease in mean arterial blood pressure to below 85 mmHg or required specific treatment for hypotension. The rates of mechanical ventilation, myocardial infarction and increase in O_2-saturation in these patients were similar to the rates in the rest of their groups.

The main outcome measures are presented in Figure 2. The composite endpoint (i.e. one or more of the three main outcome measures: death, mechanical ventilation or myocardial infarction) was recorded in 13 (25%) patients in group A and in 24 (46%) in group B.

Secondary outcome measures: changes in O_2-saturation are given in Figure 3. The heart rate decreased from 117 ± 18 to 102 ± 15 beats/min in group A ($p < 0.001$) as compared to a decrease from 113 ± 22 to 104 ± 19 beats/min in group B ($p < 0.001$). The respiratory rate decreased from 42 ± 17 to 31 ± 14 breaths/min in group A ($p = 1*10-13$) as compared to a decrease from 35 ± 8 to 30 ± 8 breaths/min in group B ($p < 0.001$).

Significant relief of congestive symptoms was achieved in both treatment groups. However, the effect of treatment in group A was dramatically superior to that of treatment in group B. The regimen applied in group B, including furosemide, morphine and low-dose nitrates, is the classic approach to the treatment of pulmonary oedema. This treatment combination causes mainly venodilatation and therefore reduction of preload.

The diuretic effect of furosemide starts only 30 minutes after administration and peaks at 1–2 hours. As the effects of these treatments were evaluated mainly during the first 60 minutes, we believe that the contribution of diuresis was not significant.

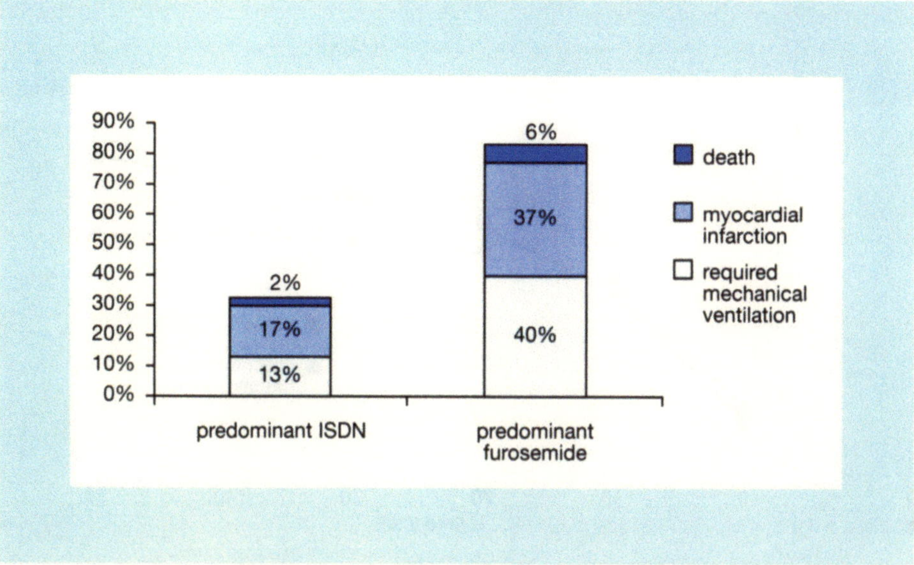

Figure 2: Results for primary outcome measures (death, myocardial infarction, requirement of mechanical ventilation) in patients treated predominantly with ISDN (left) and those treated predominantly with furosemide (right) (3).

High-dose ISDN causes arteriodilation, reducing systemic peripheral resistance and thus also afterload (4). The decrease in afterload, by increasing cardiac output, relieves the acute heart failure and improves pulmonary congestion (5). Furthermore, nitrates, by directly improving myocardial relaxation, may reduce LVEDP and abort oedema. Accordingly, we suggest that high-dose nitrates, by causing a reduction in both preload and afterload, may confer better relief of severe pulmonary oedema than furosemide (6). The findings of this study substantiate this hypothesis.

In the present study, the rate of myocardial infarction showed a significant decrease of 53% in the group of patients treated with high-dose nitrates. This may be related to

1. direct effect of high-dose nitrates on ischaemia relief (through arteriodilation)
2. indirect effect through more rapid control of pulmonary oedema, hence improved oxygenation.

Cardiogenic shock

In contrast to pulmonary oedema we suggest that in cardiogenic shock the imbalance of vascular mediators leads to a relatively suboptimal peripheral vasoconstric-

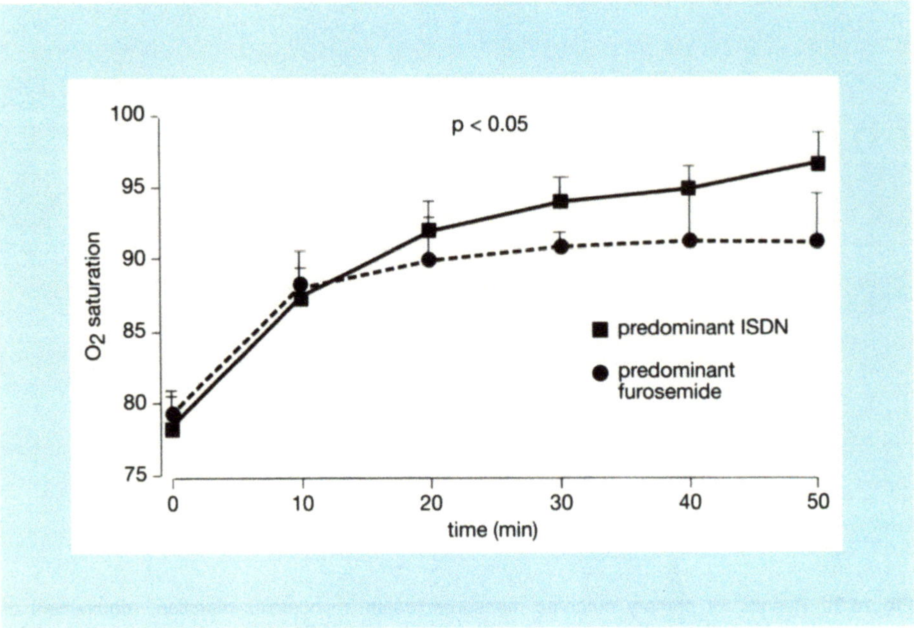

Figure 3: Pulmonary oedema: change in oxygen saturation during treatment in patients treated predominantly with ISDN and those treated predominantly with furosemide.

tion (Figure 4) (7). This in turn causes persistent vital organ hypoperfusion including decreased perfusion pressure to coronaries. The decreased coronary flow deteriorates myocardial performance and cardiac output. In addition it is possible that high levels of NO cause global myocardial dysfunction, which may further deteriorate contractility and output. Therefore, the use of a NO synthase inhibitor (L-NMMA) may restore this imbalance, enabling an increase in blood pressure without tachycardia, leading to restored perfusion of vital organs including the heart. This will terminate the "evil cycle" and change the course of the patient from continuing deterioration to gradual improvement. We have tested this hypothesis by administering L-NMMA (a NO synthase inhibitor) to patients with cardiogenic shock.

Patients and methods

Patients with extensive myocardial infarction complicated with cardiogenic shock were considered for this study. Immediately upon admission, all patients underwent coronary catheterisation and primary percutaneous transluminal coronary revascularisation (PTCR) whenever feasible. An intra-Aortic Balloon Pump (IABP) was inserted during coronary catheterisation or immediately prior to it. All patients were treated by mechanical ventilation and high doses of inotropic cathecholamines.

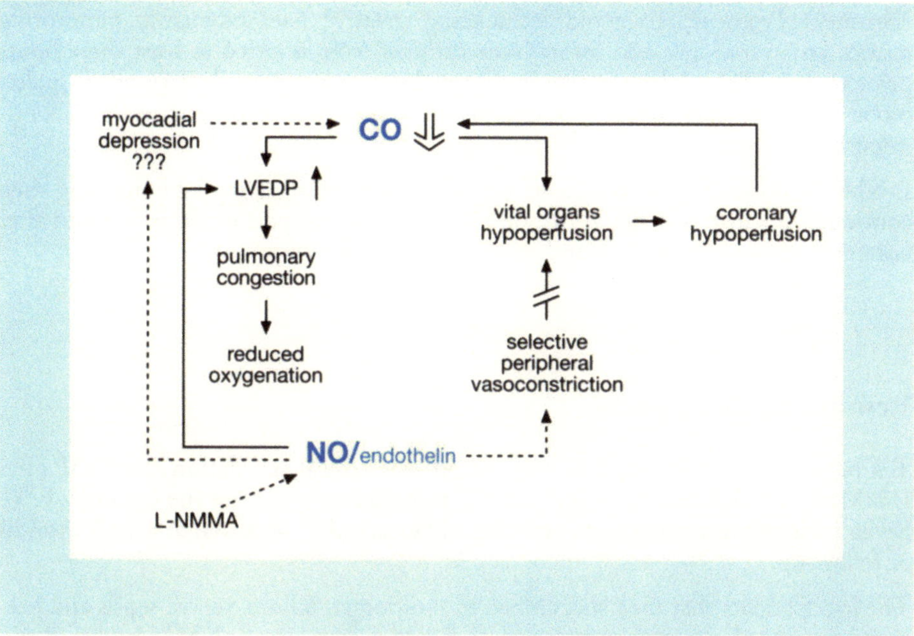

Figure 4: Cardiogenic shock: a suboptimal peripheral vasoconstrictive reaction in cardiogenic shock causes persistent vital organ hypoperfusion including coronary hypoperfusion. The decreased coronary flow deteriorates myocardial performance and cardiac output (CO).

Cardiogenic shock was defined as persistent unaugmented systolic blood pressure (systolic blood pressure measured without IABP augmentation) below 100 mmHg, accompanied by pulmonary congestion determined by chest X-ray, cardiac index < 2.5 $1/min./m^2$ and wedge pressure > 15 mmHg despite the above mentioned treatment.

Concomitant treatment: All patients received aspirin, i.v. heparin, i.v. fluids, i.v. furosemide drip, i.v. dopamine and i.v. dobutamine. Dopamine and dobutamine were administered to doses of at least 10 microgram/kg/minute for at least 3 hours prior to enrolment. During L-NMMA administration, treatment with fluids, cathecholamines, mechanical ventilation and IABP was kept constant.

Inclusion and exclusion criteria: Patients were enrolled if having refractory cardiogenic shock (cardiogenic shock persisting or worsening > 24 hours after admission and coronary catheterisation) and if deemed beyond treatment by the heads of cardiology and coronary intensive care unit (CICU). Subsequent to patients' selection and approval by the heads of cardiology and CICU, patients and families were required to sign an informed consent form. Patients with significant tachy- or bradyarrhythmias, significant valvular heart disease or other mechanical complications, secondary heart failure, sepsis (determined by fever > 38 degrees), or creatinine > 200 micromoles/ml, were excluded.

Treatment Protocol: To avoid influencing invasive haemodynamic monitoring results, an arterial line and Swan-Ganz catheter were inserted at least three hours prior to L-NMMA administration. Throughout treatment, O_2-saturation, pulse, blood pressure, urine output, wedge pressure, and cardiac output were all continuously monitored.

L-NMMA (Clinalfa, Cal-biochem) was administered as a 1 mg/kg bolus, and then continued as a drip of 1 mg/kg/hr for 5 hours. ECG was performed before and after administration of L-NMMA and at 24 hours of follow-up.

Results and discussion

Ten out of 11 patients could be weaned off mechanical ventilation and IABP after L-NMMA administration. Eight patients were discharged from the coronary ICU. Seven patients were discharged to home. They are alive and well at 1 to 3 months of follow-up.

The four patients that died succumbed to multiorgan failure, sepsis, sepsis and haemorrhage and cholesterol emboli at 1, 2, 3 and 6 days after L-NMMA administration, respectively.

Changes in mean arterial blood pressure (MAP) urine output, cardiac index and pulmonary capillary wedge pressure are shown in Figure 5. Heart rate decreased by 6% after 10 minutes of treatment and remained largely unchanged during L-NMMA administration and at 24 hours follow-up. No patients died during L-NMMA administration. We were unable to detect any clinical or laboratory adverse effects of L-NMMA treatment.

Two possible theories could explain the results of the present study:

1. A haemodynamic effect of L-NMMA:

It is clear from the haemodynamic monitoring we have performed that L-NMMA is a potent, selective, peripheral vasoconstrictor. Its administration has resulted in a 76% increase in SVR and a 41% increase in MAP. This induces an immediate increase in afterload, therefore explaining the initial decrease in cardiac index and minor increase in pulmonary capillary wedge pressure. However, the decrease in cardiac index and increase in wedge pressure were transient. Starting at 3 hours from L-NMMA administration, the cardiac index gradually increased despite a persistent elevated SVR and MAP, which implies, as noted previously, that cardiac contractility was improved at this stage. This late regain in cardiac index could be explained by improved myocardial perfusion. Such effect of reduced blood pressure on myocardial perfusion and patient outcome could also explain the results of the CONSENSUS-II-study, where early decrease in blood pressure by an ACE inhibitor was detrimental to patients with acute MI (8).

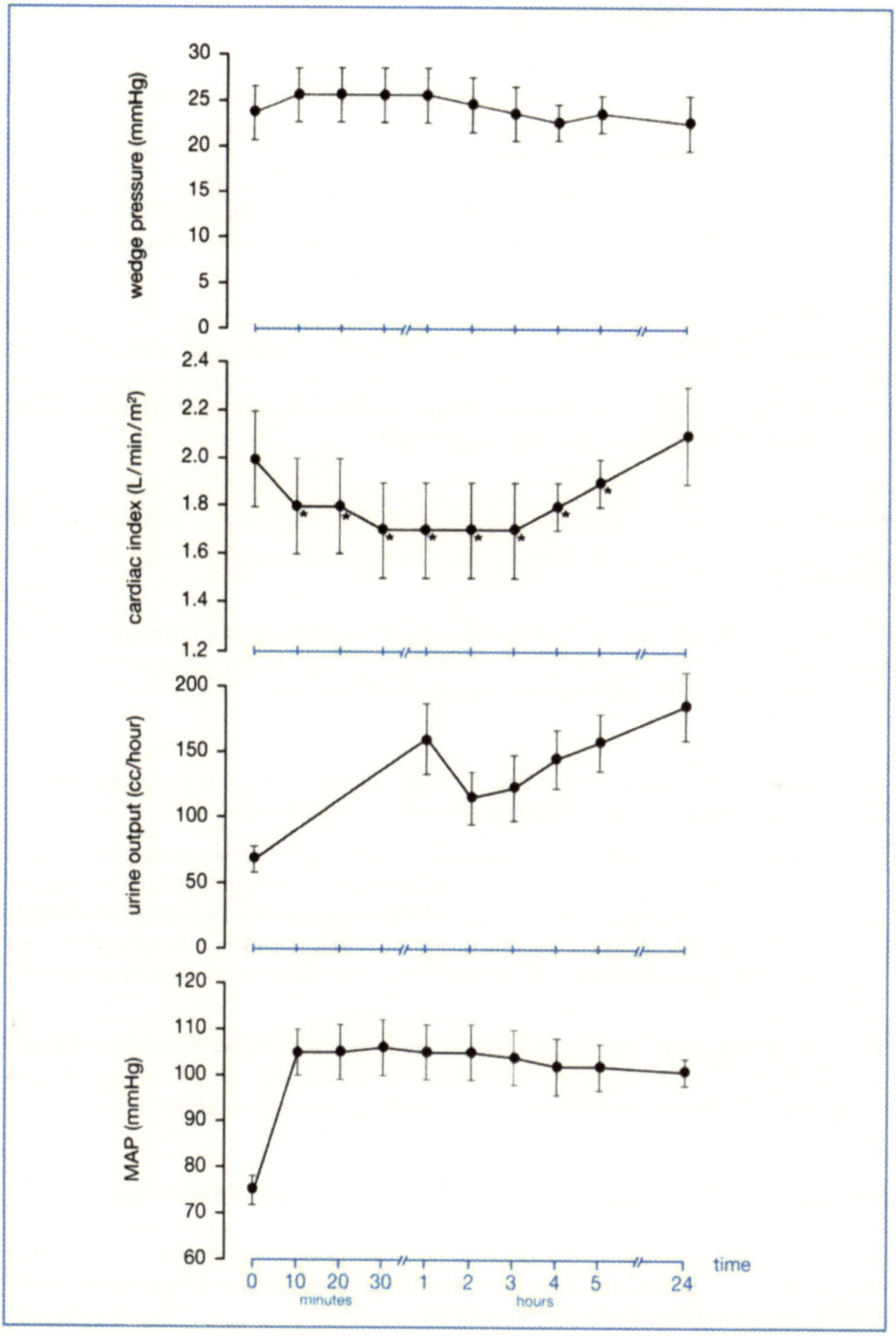

Figure 5: Cardiogenic shock: changes in mean arterial blood pressure (MAP), urine output, cardiac index, and pulmonary artery wedge pressure (PAWP) after L-NMMA application.

2. Direct effect on myocardial contractility:

As stated by H. Drexler in a recent editorial (9), evidence is accumulating that NO in the failing human heart is a double-edged sword. Many research groups have reported that NO, especially when present at a high concentration, is a cardiodepressant. The mechanism of action of this effect of NO is not clear. However, it is possible that during shock high NO levels induce a cardiodepressant effect. Thus, some of the clinical improvement we observed starting at 3 hours after L-NMMA administration and persisting throughout the 24 hours of invasive haemodynamic monitoring could be explained by inhibition of the direct effect of NO on the myocardium.

Therefore, L-NMMA administration has favourable clinical and haemodynamic effects on patients with cardiogenic shock. Further studies are needed to determine the dose-response curve of L-NMMA and mechanism of action, particularly in relation to ventricular systolic and diastolic function.

Summary

The results of our studies indicate that during an acute decrease in cardiac function, and hence stroke volume, three scenarios are possible regarding the reaction of vascular control and its interaction with cardiac performance:

1. Adequate vasoconstriction leading to adequate redistribution of cardiac output and achievement of a new balance of supply and demand.
2. An excessive vasoconstrictive reaction leading to a steep increase in afterload, initiating a viscous cycle terminating in pulmonary oedema. In this setting we were able to demonstrate that the use of intravenous (i.v.) boluses of Isosorbide Dinitrate (ISDN) reduces congestion significantly, improving oxygenation, and reducing the need for mechanical ventilation and the rate of myocardial infarctions.
3. Too little vasoconstriction: In a recent preliminary study we demonstrated that reducing NO production by L-NMMA in patients with post MI, refractory cardiogenic shock increases blood pressure and improves clinical outcome.

Furthermore, in acute ischaemic syndromes our recent study showed that the use of intravenous boluses of Isosorbide Dinitrate to reduce mean arterial blood pressure by 5–20% improves ischaemia control. In conclusion, our results indicate that manipulation of the NO pathway significantly affects vascular control and can be used in the treatment of acute cardiac conditions.

References

1. Cotter G, Faibel H, Barash P, et al. High-dose nitrates in the immediate management of unstable angina: optimal dosage, route of administration, and therapeutic goals. Am J Emerg Med 1998; 16: 219–224.
2. Morrison RA, Weigand UW, Jähnchen E, et al. Isosorbide dinitrate kinetics and dynamics after intravenous, sublingual and percutaneous dosing in angina. Clin Pharmacol Ther 1983; 33: 747–756.
3. Cotter G, Metzkor E, Kaluski E, et al. Randomized trial of high-dose isosorbide dinitrate plus low-dose furosemide versus high-dose furosemide plus low-dose isosorbide dinitrate in severe pulmonary oedema. Lancet 1998; 351: 389–393.
4. Imhof PR, Ott B, Frankhauser P, et al. Difference in nitroglycerin dose response in the venous and arterial beds. Eur J Clin Pharmacol 1980; 18: 455–460.
5. Nelson GIC, Silke B, Ahuja RC, et al. Haemodynamic advantages of isosorbide dinitrate over furosemide in acute heart failure following myocardial infarction. Lancet 1983; 730–732.
6. Northridge D. Frusemide or nitrates for acute heart failure? Lancet 1996; 347: 667–668.
7. Cotter G, Kaluski E, Blatt A, et al. L-NMMA (a nitric oxide synthase inhibitor) is effective in the treatment of cardiogenic shock. Circulation (in press).
8. Swedburg K, Held P, Kjekshus J, et al. Effects of early administration on enalapril on mortality in patients with acute myocardial infarction. Results of the cooperative north Scandinavian enalapril survival study II (CONSENSUS II). N Engl J Med 1992; 327: 678–684.
9. Drexler H. Nitric oxide synthases in the failing heart: a double-edged sword? Circulation 1999; 99: 2972–2975.

Economic impact of Elantan LA compared to Tenormin and Tildiem LA in the treatment of stable angina in the UK

T. Baxter, A. Rehe

Introduction

Stable angina usually reflects inadequate blood supply in the coronary arteries following the development of an atherosclerotic plaque. As a result, most cases of stable angina are manifestations of CHD, the leading cause of mortality in the UK and a leading cause of morbidity (1–3). In 1994, the annual cost of managing CHD was estimated to be £60 million, accounting for 2.5% of NHS expenditure (4).

The overall prevalence of angina in the population over 30 years of age is 24,000 cases per million (average of men and women) with about 800 new cases per million population per year (5). In a year, 1% of the population present with anginal symptoms to a GP (6). Of these, about 10% will either have a non-fatal MI or die from coronary causes (7).

There are three main management strategies for angina: symptomatic and prophylactic medical treatment, percutaneous transluminal coronary angioplasty (PTCA), and coronary artery bypass grafting (CABG). Additionally, interventions should be accompanied by risk factor modification, such as smoking cessation, increased exercise, weight reduction and dietary change (8). Medical treatment of stable angina involves three classes of drugs; nitrates; betablockers; and calcium antagonists. The few published comparisons showed no major difference in the ability of these classes to relieve angina (9–15). Furthermore, there is no evidence that combination therapy is more effective than monotherapy (15–17).

Nitrates are potent venodilators and, to a lesser extent, arterial dilators. However, their main clinical benefit arises from peripheral vasodilation. As a result, nitrates reduce cardiac workload. Nitrates are safe with concurrent calcium antagonists and betablockers and are safe when lung function is impaired – unlike many betablockers (18, 19). Betablockers reduce sympathetically mediated effects on myocardial demand by reducing heart rate, blood pressure and contractility. Cardio-selective betablockers preferentially inhibit cardiac beta-1-receptors. However, cardio-selectivity decreases with increasing dose. For example, atenolol 100 mg is less cardio-selective than 50 mg. Nevertheless, atenolol is the preferred betablocker in angina: it is cardio-selective and hydrophilic (18, 19). Calcium antagonists impair the influx of calcium ions into smooth muscle, myocardial and conducting tissue cells. Hence, calcium antagonists reduce myocaridial contractility and may also depress conduction and relax coronary and peripheral vascular tone. In general, calcium antagonists are used when betablockers are either contraindicated or have induced side-effects (5). Verapamil and diltiazem are the main calcium antagonists used as alternatives to betablockers in angina (3). Due to the risks of bradycardia, heart block and heart failure, these agents should only be administered with betablockers under caution.

Many studies have examined the costs and cost-effectiveness of PTCA and CABG. However, few studies have compared the costs and cost-effectiveness of medical treatments. Against this background we used the IMS longitudinal database, Mediplus to estimate the economic impact to the UK's NHS of treating new, switched and existing stable patients with Elantan LA for one year compared to Ternormin and Tildiem LA.

Methodology

Perspective

This evaluation estimated the direct healthcare costs over one year of managing new, switched and existing stable angina patients from the perspective of the UK's NHS.

NHS resource use and costs

The longitudinal IMS database, Mediplus contains information on one million patients from 1,117 GPs in 153 primary care practices across the UK. In the database, the most commonly prescribed betablockers and calcium antagonists for stable angina was atenolol (Ternomin) and long-acting diltiazem (Tildiem LA) respectively. Consequently, Mediplus was interrogated to obtain angina-related healthcare resource use over one year for patients who received either Elantan LA or one of the two other drugs for stable angina and had one of the following Read Codes: history of angina pectoris; history of angina in the last year; angina control; and angina pectoris.

Three patient groups were studied:

1. New patients who were diagnosed and treated for angina during 1995 or 1996. They were tracked for 12 months from first treatment.
2. Switched patients who were diagnosed as suffering from angina for a median 4.5 years and who were switched from their existing therapy during 1995 or 1996. They were tracked for 12 months from the switch.
3. Existing patients: who were diagnosed as suffering from angina for a median 4.5 years and who received continuous treatment with one of the study drugs for at least one year before the study period. They were tracked for 12 months from 1 January 1996.

Mediplus provided resource use estimates over one year on the number of GP visits, GP prescriptions, tests initiated by GPs, practice nurse visits, outpatient visits to hospital specialists, elective and emergency hospitalisations that included blood cell transfusions, cardiac cathetherizations, continuous ECG monitoring, coronary artery operations, direct gastroscopies and endoscopies.

Resource use was stratified according to the following comorbidities; ischaemic heart disease (excluding angina), hypertension, congestive cardiac failure, hypercholesterolaemia and cerebrovascular disease.

Time between prescription issues for the anti-anginals, taking into account prescription size and dosing instruction, was used as a proxy for compliance. One hundred percent compliance was considered to have occurred when the elapsed time between successive prescriptions was exactly equal to the theoretical duration of the earlier prescription.

Unit resource costs were obtained from published literature (20–24) and NHS databases (25–26) and standardised to 1997/98 prices. The acquisition cost of the branded anti-anginals was used irrespective of whether patients received a branded or generic product since resource use attributable to a specific drug is likely to be the same irrespective of the drug dispensed. By assigning unit costs to the estimates of resource use, the cost to the NHS of managing angina patients with Elantan LA, Tenormin and Tildiem LA for one year was estimated.

Sensitivity analyses

A sensitivity analysis tests a study's robustness by varying key parameters that either strongly influence the results or are based on uncertain assumptions. As a result, a sensitivity analysis assess this study's robustness by varying the frequency of GP visits and the incidence of hospitalisation.

Results

New patients

Compliance with each anti-anginal was 100 % across all treatments. There were no significant differences in resource use between new patients who received the same anti-anginal, when stratified by comorbidity. There was a difference between treatment groups in the incidence of elective hospitalisation irrespective of comorbidity as was the case with the number of practice nurse visits. Differences in the incidence of emergency hospitalisations among new patients with comorbidity were also significant.

Overall, the incidence of hospitalisation among new Tenormin-treated patients with comorbidity and new Tildiem LA-treated patients without comorbidity was at least three times higher than in the other groups (Figure 1). Accordingly, the cost of treating these new patients in their first year of anti-anginal treatment was at least £300 more than the cost of new patients in the other groups.

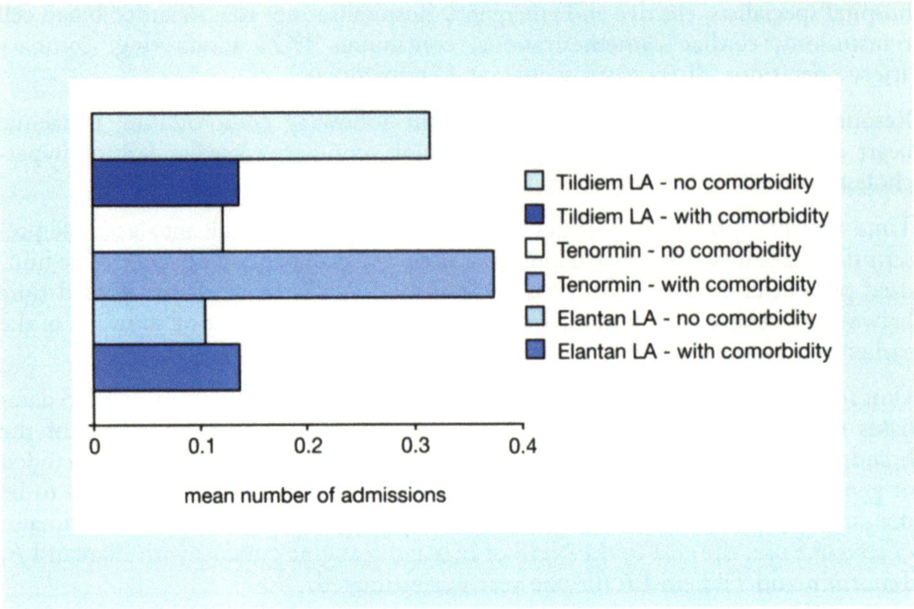

Figure 1: New patients – hospital admissions.

The cost of managing a new patient in the community was largely unaffected by comorbidity, and was similar across treatment groups (range £270 to £355). In contrast, comorbidity affected secondary care costs. The difference was particularly striking among new Tenormin-treated patients since patients with comorbidity cost over 2.5 times more to manage than those without comorbidity (£372 versus £985). Similarly, new Tildiem LA-treated patients without comorbidity cost over twice as much to manage as those with comorbidity (£345 versus £740). These differences in secondary care costs may reflect the choice of anti-anginal treatment.

Switched patients

Compliance with each anti-anginal was again 100% across all treatments. There were no differences in resource use between switched patients who received the same anti-anginal, when stratified by co-morbidity. However, there were marked differences between treatment groups in the incidence of emergency hospitalisation irrespective of comorbidity. There were also differences between the groups in the incidence of elective hospitalisation among switched patients without comorbidity (Figure 2). The incidence of hospitalisation among patients with comorbidity who were switched to Elantan LA was at least 2.5 times lower than among patients who were switched to one of the other anti-anginals. None of the patients without comorbidity who were switched to Elantan LA were hospitalised, compared to 20% of patients who were switched to Tenormin and Tildiem LA.

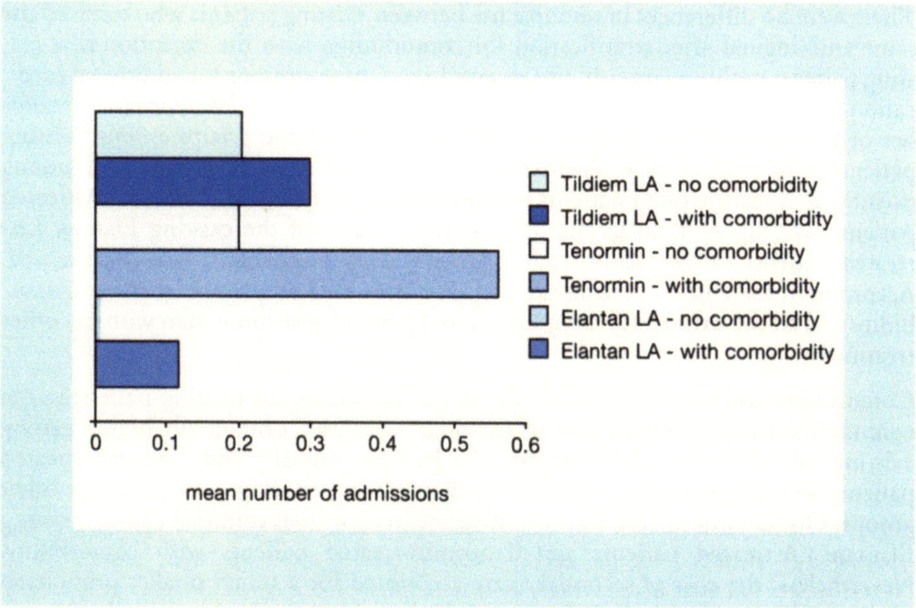

Figure 2: Switched patients – hospital admissions.

Additionally, the number of GP-initiated tests, practice nurse visits and outpatient visits was significantly different among patients with comorbidity.

The total cost of a patient with comorbidity during the first year after switching to Elantan LA was at least £400 less than for patients switched to one of the other anti-anginals. Furthermore, the cost of treating an Elantan LA-treated patient without comorbidity in the first year after switching was £400 less than the cost of switching a patient to Tenormin and Tildiem LA.

The cost of managing a switched patient in the community was unaffected by comorbidity, and was approximately similar across treatment (£218 to £372). In contrast, the cost of secondary care was higher among patients who had comorbidity, irrespective of their anti-anginal. Overall, secondary care costs were higher among patients switched to Tenormin (£1,108) and lowest among patients switched to Elantan LA (£336).

Existing patients

Existing patients had taken the same anti-anginal for at least a year before being tracked in the study. Compliance was 100% during the year before the study period and 93% during the study period across all groups.

There were no differences in resource use between existing patients who received the same anti-anginal after stratification for comorbidity, with the exception that existing patients without comorbidity received more prescriptions for additional medication than those with comorbidity. Additionally, there was a difference in the number of GP-initiated tests and outpatient visits between the groups among existing patients with comorbidity. Furthermore, the incidence of hospitalisation among existing Tenormin-treated patients without comorbidity and all Tildiem LA-treated patients was higher than in the other groups. None of the existing Elantan LA-treated patients or Tenormin-treated patients with comorbidity was hospitalised. Accordingly, the cost of treating existing Tenormin-treated patients without comorbidity and all the existing Tildiem LA-treated patients was more than with the other treatments.

Comorbidity had little impact on the cost of managing an existing patient in the community. In contrast, the cost of secondary care was substantially higher among existing Tildiem LA-treated patients who had comorbidity and Tenormin-treated patients who did not have comorbidity than among existing patients in the other groups. On the other hand, the cost of secondary care was minimal among existing Elantan LA-treated patients and Tenormin-treated patients with comorbidity. Nevertheless, the cost of secondary care accounted for a much smaller proportion of the total cost compared to the new and switched patients. Accordingly, the primary care cost was the main cost driver in the management of existing patients.

Sensitivity analyses

In addition to the acquisition cost of the anti-anginals, hospitalisation and GP visits were the main cost drivers. Therefore, sensitivity analyses determined how varying the incidence of hospitalisation and the frequency of GP visits would influence the cost of each treatment strategy.

New patients

GP visits accounted for between 11% and 16% of the total cost of managing angina during the first year of treatment. However, a sensitivity analysis showed the results to be robust to changes in the frequency of GP visits.

Secondary care resource use accounted for between 53% and 71% of the total cost of managing angina during the first year of treatment. However, a sensitivity analysis showed that the cost per Elantan LA-treated patient remains lower than that of either a Tenormin- or Tildiem LA-treated patient, except when no patients are hospitalised in any group.

Switched patients

GP visits accounted for between 11 % and 21 % of the total cost of managing a patient during the first year after switching treatment. However, a sensitivity analysis showed the results to be robust to changes in the frequency of GP visits.

Secondary care resource use accounted for 37 % of the total cost of managing a patient during the first year after switching to Elantan LA. This compared to between 60 % and 75 % with each of the other anti-anginals. A sensitivity analysis showed that the cost per Tenormin-treated patient remains higher than with the other anti-anginals, except when no patients are hospitalised in any group.

Existing patients

GP visits accounted for between 22 % and 41 % of the total cost of managing an existing patient during the study period. However, a sensitivity analysis showed that the results are robust to changes in the frequency of GP visits. However, the cost per patient in the Elantan LA and Tenormin groups crossover when the frequency of GP visits is three times above baseline for both treatments.

Secondary care resource use accounted for 1 % of the total cost of managing an existing patient with Elantan LA compared to between 26 % and 40 % with the other anti-anginals. However, a sensitivity analysis showed that the results are robust to changes in the incidence of hospitalisation from zero to double the baseline value, although the cost per Tenormin-treated patient falls below that of a patient treated with Elantan LA when the incidence of hospitalisation in the Tenormin group is reduced by about one third. None of the Elantan LA-treated patients were hospitalised.

Discussion

Numerous studies have examined the economic impact of PTCA and CABG. However, few studies have compared the economic impact of different drug classes in the treatment of stable angina. The analysis showed that primary care resource use was broadly similar across the drug classes and that the cost differential between treatments was largely attributable to differences in the incidence of hospitalisation. These differences may be an artifact of the Mediplus dataset, since Mediplus is principally a primary care database and the GP panel may have underreported secondary care resource use. However, such underreporting would apply equally to all three patient types in all groups. Alternatively, the differences in hospitalisation may be a direct or indirect effect of the anti-anginals and should be investigated further in a prospective study. Nevertheless, sensitivity analyses showed that the results were robust to realistic changes in the incidence of hospitalisation as well as the frequency of GP visits.

The analysis also showed that the prescribing of additional medication and other resource use seems unaffected by comorbidity, except in existing patients without comorbidity who received significantly more prescriptions than existing patients with comorbidity. This finding reflects clinical practice as recorded by Mediplus, although the reason seems unclear. There was widespread use of aspirin, lipid lowering drugs and diuretics, which were prescribed presumably for secondary prevention of cardiac events (5, 18, 27, 28).

The costs in our study (Figure 3) are comparable to those estimated by O'Neil et al. (29), who investigated the cost-effectiveness of personal health education for angina patients being treated in general practice in Northern Ireland. The annual cost of healthcare resource use was around £1,800. Procedures (including hospitalisation and GP visits) accounted for 97% of the cost and drugs accounted for between £194 and £260 per patient. In this study, the cost of the anti-anginals and additional medication was approximately £200 per year. Additionally, O'Neil et al. found significant improvements in survival and self-assessed quality of life among patients who received health education compared to controls, although there were no significant differences in the cost of resource use.

Another study compared the costs and outcomes of once-daily isosorbide mononitrate (ISMN) with twice daily ISMN and transdermal nitrate (30). The annual acquisition cost of once-daily ISMN was £149 per patient compared to £60 for the twice-daily drug and £201 for the transdermal patch. However, patients treated with twice daily ISMN used more healthcare resources and had a lower compliance rate than patients who received either of the other treatments. Accordingly, the annual cost of treating exercise-induced angina with once-daily ISMN was £248 per patient, compared to £250 for a patient treated with twice-daily ISMN and £299 with the transdermal patch. However, the resource use estimates in this study were

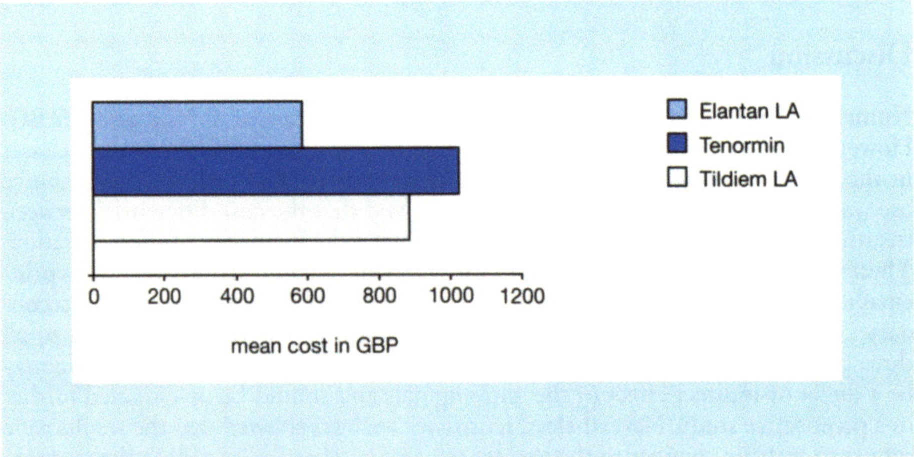

Figure 3: Total resource cost.

substantially lower than that of patients in the Mediplus dataset. Consequently, this study (30) appears to have substantially underestimated the actual costs of managing angina patients.

In a study comparing the cost-effectiveness of medical treatment relative to CABG (31), the cost per life-year gained (cost per QALY in parenthesis) over five years with CABG versus initial standard medical therapy and aspirin was estimated to be US $73,601 ($36,709) (31). This compared to US $229,077 ($55,156) when a statin was included. In contrast, the cost per life-year gained over five years with initial standard medical therapy plus aspirin and a statin versus initial standard medical therapy plus aspirin was US $34,883 (£23,730). The cost-effectiveness of surgery relative to drugs improved with the severity of angina, since the costs per life-year gained a per QALY decreased with increasing angina severity.

Some cardiologists believe CABG benefits only patients with either three-vessel coronary or left main stem disease, whereas others consider that only patients with left ventricular dysfunction benefit. Moreover, other cardiologists suggest that CABG should be reserved to improve the prognosis of younger patients. However, based on the data from this study and the views of others (31) drugs control most patients with stable angina most of the time. Drugs offer the greatest efficacy for most patients, since invasive procedures carry a risk of morbidity and mortality (32–35). These risks need to be weighed against PTCA's and CABG's advantages, such as reduced risk of infarction or death and fewer symptoms (32, 35–37). But stable angina's prognosis when treated medically is improving (31), particularly with the widespread use of aspirin. Moreover, further improvements are expected with more aggressive medical management (5). Nevertheless, for patients who develop intractable angina, drugs alone are no longer satisfactory and PTCA or CABG becomes necessary.

One of the limitations of using a longitudinal database is the need to assume that the clinical effectiveness of all the interventions under investigation are similar. Nevertheless, some recent treatment guidelines were developed on this basis, using drug acquisition costs as the only criteria for using an anti-anginal (38). Accordingly, the data can be used to draw some inferences about suitable anti-anginal treatments for specific patients. Accordingly, policy formers and decision makers should consider the broader resource use implications of choosing an anti-anginal rather than drug acquisition costs alone when developing treatment guidelines. Furthermore, the actual financial impact of a treatment strategy will be determined by the resource use attributable to that treatment and not exclusively on drug acquisition costs.

This analysis has several limitations. The analysis was limited to those patients who remained on the same treatment for at least one year in the Mediplus database. The analysis did not consider patients who died or were lost to follow-up. Neither did the study evaluate differences in clinical effectiveness between particular drugs. Also, the analysis did not consider the clinical consequences of angina, such as MI or death, as a result of limitations in the Mediplus database. However, the drug classes examined in this study are equally effective in the treatment of angina (9–16) and there is no evidence to suggest that the incidence of such outcomes would vary between them.

Patients were not randomised and GP's decision to prescribe a particular anti-anginal may be based on past experience, formularies or other clinical and non-clinical factors. However, the patient data contained within the Mediplus database reflects actual medical practice on a large number of patients without the selection biases that may arise in intervention studies. Additionally, as a result of limitations in the Mediplus database, statistical tests were based on differences in resource use by patients and not on differences in GP's behaviour, such as the use of a specific test. However, any such bias should be minimised since the dataset was derived from a sufficiently larger number of GPs (n = 1, 117) and primary care practices (n = 153). Furthermore, the distribution of patients according to age and sex is representative of the UK. Moreover, the economic comparisons were retrospective and the distribution of potentially influential variables – such as duration of disease and other comorbidities – were assumed to represent the angina population.

This analysis provides an economic snapshot of stable angina management by GPs and provides a framework that can be refined and extended as more data become available. The analysis excluded direct costs to patients and their families, indirect costs to society, such as loss of productivity, and intangible costs, such as changes in quality of life.

In conclusion, clinical decisions about choosing between Elantan LA, Tenormin and Tildiem LA, which all have similar effectiveness, should not be based on their acquisition cost alone. By accounting for a broader range of healthcare resource use, the nitrates were found to generate economic benefits in all patient groups.

Summary

The study estimated the cost to the UK's National Health Service (NHS) of treating stable angina with Elantan LA (long-acting isosorbide mononitrate 50 mg) compared to Tenormin (atenolol) and Tildiem LA (long-acting dilitiazem). Angina-related healthcare resource use from IMS database, Mediplus was obtained for new and switched patients who received one of the three drugs for stable angina and who also had a stable angina-related Read Code. Resource use was stratified according to a patient's comorbidity. National resource costs (at 1997/98 prices) were assigned to the resources used, to estimate the cost of each treatment strategy over one year to achieve the same level of clinical effectivenes.

The cost differences between treatments were largely attributable to differences in hospitalisation rates, since primary care resource use was broadly similar across treatments. Total hospitalisation rates among new patients with comorbidity who were treated with Tenormin and new patients without comorbidity who were treated with Tildiem LA, were at least 3 times higher than amongst the new patients in the Elantan LA group (with or without comorbidity).

The incidence of hospitalisation among switched patients with comorbidity who received Elantan LA was at least 2.5 times lower than in the other groups. None of the patients without comorbidity who were switched to Elantan LA were hospitalised. In contrast, 20% of patients who were switched to Tenormin and Tildiem LA were hospitalised and this was reflected in the higher costs.

In conclusion, clinical decisions about choosing between Elantan LA, Tenormin and Tildiem LA, which all have similar effectiveness, should not be based on their acquisition cost alone. By accounting for a broader rang of healthcare resource use, Elantan LA was found to generate economic benefits in all patient groups.

References

1. Office of Health Economics, compendium of Health Statistics 1992. London: Office of Health Economics 1993.
2. Shaper AG, Cook DG, Walker M, et al. Prevalence of ischaemic heart disease in middle aged British men. Br Heart J 1984; 51: 595–605.
3. White A, Nicholaas G, Foster K, et al. Health survey for England 1991. HMSO London 1993.
4. Gunnell D, Harvey I, Smith L. The invasive management of angina: issues for consumers and commissioners. J Epidemiol Community Health 1995; 49: 335–343.
5. Cleland JGF. Can improved quality of care reduce the costs of managing angina pectoris? Eur Heart J 1996; 17 (suppl A): 29–40.
6. McCormick A, Fleming D, Charlton J. Morbidity statistics from general practice: fourth national study 1991–1992, HMSO; London, 1995.
7. Ghandhi MM, Lampe FC, Wood DA. Incidence, clinical characteristics, and short-term prognosis of angina pectoris. Br Heart J 1995; 73: 193–198.
8. North of England stable angina guidelines development group, North of England evidence based guidelines development project: summary version of evidence based guideline for primary care management of stable angina. Br Med J 1996, 312: 827–832.
9. Destors JM, Boisell JP, Philippon AM, et al. Controlled clinical trial of bepridil, propranolol and placebo in the treatment of exercise induces angina pectoris. Fundam Clin Pharmacol 1989; 3: 597–611.
10. Vliegen HW, van der Wall EE, Niemeyer MG, et al. Long-term efficacy of diltiazem controlled release versus metoprolol in patients with stable angina pectoris. J Cardiovasc Pharmacol 1991; 18: 55–60.
11. Nahrendorf W, Rading A, Steinig G, et al. A comparison of carvedilol with a combination of propranolol and isosorbide dinitrate in the chronic treatment of stable angina. Cardiovasc Pharmacol 1992; 19: 114–116.
12. Boberg J, Larsen FF, Pehrsson SK. The effects of beta blockade with (epanolol) and without (atenolol) on intrinsic sympathomimetic activity in stable angina pectoris, Clin Cardiol 1992; 15: 591–595.
13. Singh S. long-term double-blind evaluation of amlodipine and nadolol in patients with stable exertional angina pectoris. Clin Cardiol 1993; 16: 54–58.
14. Guermonprez JL, Blin P, Peterlongo F. A double-blind comparison of long-term efficacy of a potassium chennel opener and a calcium antagonist in stable angina pectoris. Eur Heart J 1993; 14: 30–34.

15. Dargie HJ, Ford I, Fox KM. Total Ischaemic Burden European Trial (TIBET). Effects of ischaemic treatment with atenolol, nifedipine SR and their combination on outcome in patients with chronic stable angina. Eur Heart J 1996; 17: 76–81.
16. Rehnqvist N, Hjemdahl P, Billin E, et al. Effects of Metoprolol vs verapamil in patients with stable angina pectoris. The Angina Prognosis Study in Stockholm (APSIS). Eur Heart J 1996; 17: 76–81.
17. Savonitto S, Ardissiono D, Egstrup K, et al. Combination therapy with metroprolol and nifedipine versus monotherapy in patients with stable angina pectoris. Results of the International Multicenter Angina Exercise (IMAGE) Study. J Am Coll Cardiol 1996; 27: 311–316.
18. Cleland JGF, McMurray J, Ray S. Prevention strategies after myocardial infarction. Science Press London 1994.
19. Dollery C (ed.). Drugs and Therapeutics. Churchill-Livingstone, Edinburgh, 1991.
20. Monthly Index of Medical Specialities (MIMS) 1998 editions, Haymarket Publications Limited, London.
21. Netten A, Dennett J. Unit costs of health and social care 1997. University of Kent, Canterbury. Personal Social Services Research Unit 1997.
22. Guest JF, Munro VL. A comparison of the economic impact on the NHS of ramipril and enalipril in the treatment of essential hypertension. Br J Med Economics 1996; 10: 303–314.
23. Guest JF, Munro VL, Cookson RF. The annual cost of blood transfusion in the United Kingdom. Clinical & Laboratory Medicine 1998; 20: 111–118.
24. Guest JF, Morris A. Community-acquired lower respiratory tract infections: the annual cost to the national Health Service. Br J Med Economics 1996; 10: 263–273.
25. Chartered Institute of Public Finance and Accountancy (CIPFA): Health Database 1997, London.
26. Department of Health 1998: NHS costs of elective and non-elective inpatients.
27. Scandinavian simvastatin survival study (4S): Randomised trial of cholesterol lowering in 4,444 patients with coronary heart disease. Lancet 1994; 344: 1383–1389.
28. Standing Medical Advisory Committee. The use of statins. Department of Health, May 1997.
29. O'Neil C, Normand C, Cupples M, et al. Cost effectiveness of personal health education in primary care for people with angina in the Greater Belfast area of Northern Ireland. J Epidemiol Community Health 1996; 50: 538–540.
30. Brown RE, Kendall MJ, Halpern MT. Cost analysis of once-daily versus twice-daily ISMN or transdermal patch for nitrate prophylaxis. J Clin Pharm Therap 1997; 22: 67–76.
31. Cleland JGF, Walker A. Is medical treatment for angina the most cost-effective option? Eur Heart J 1997; 18 (suppl. B): B35–B42.
32. RITA-2 trial participants. Coronary angioplasty versus therapy for angina: the second Randomised Intervention Treatment of Angina (RITA-2) trial. Lancet 1997; 350: 461–468.
33. Schmuziger M, Christenson JT, Maurice J, et al. Reoperative myocardial revascularisation: an analysis of 458 reoperations and 2,645 single operations. Cardiovasc Surg 1994; 2: 623–629.
34. Jagal SB, Tu JV, Naylor CD. Higher in-hospital mortality in female patients following coronary artery bypass surgery: a population-based study. Provincial Adult Cardiac Care Network of Ontario. Clin Invest Med 1995; 18: 99–107.
35. Yusuf S, Zucker D, Peduzzi P, et al. Effect of coronary bypass graft surgery on survival; overview of 10-year results from randomised trial by the Coronary Artery Bypass Graft Surgery Trialists Collaberation. Lancet 1994; 344: 563–570.
36. Versci F, Gaspardone A, Tomai F, et al. A comparison of coronary artery stenting with angioplasty for isolated stenosis of the proximal left anterior descending coronary artery. N Eng J Med 1997; 336: 817–822.
37. Rogers WJ, Coggin CJ, Gersh BJ, et al. Ten-year follow-up of quality of life in patients randomised to receive medical therapy or coronary artery bypass graft surgery. Circulation 1990; 82: 1647–1658.

38. Regional Drug and Therapeutics Centre. Modified-release isosorbide mononitrate. Drug Update, 1999; 5: NHS northern and Yorkshire.

Quality of life in angina therapy: focus on the beneficial effects of nitrates

M.G. Niemeyer, R.M.G. Jansen, T.JM. Cleophas, A.H. Zwinderman

Introduction

Measurement of quality of life (QOL) as an indicator of health outcome has become increasingly important for patients with coronary artery disease (CAD), where the goal of treatment is not only to improve prognosis, but also to relieve symptoms and improve function (1, 2). Recent research has demonstrated that variables other than disease symptoms and pain are more important to patients for their quality of life. In particular, areas such as patients' physical performance and psychological and social functioning have been recognized as important determinants. Also burden of symptoms and perceptions of well-being must be considered. Nitrates can effectively control symptoms of angina pectoris and have been demonstrated to reduce mortality in patients with acute myocardial infarction (3, 4). Based on the results of the GISSI-3 (5) and the ISIS-4 (6) studies, it is clear that nitrates can be safely used for long-term treatment. However, little is known about their net effects on current health-related QOL indices. Nitrates can be effectively applied for controlling symptoms of angina pectoris not only in patients with significant coronary artery disease but also in patients with other forms of ischaemia, such as variant angina (7) and microvascular dysfunction (8). They have also been demonstrated to increase angina-free walking (9) and to decrease exercise-induced ischaemia (10). A major problem with chronic treatment is nitrate tolerance which is probably caused by GTP-depletion in the vascular smooth muscle cells (11), rather than some sort of receptor-mediated mechanism. Although it soon develops during continued therapy (limiting the usefulness of continued therapy), it can be prevented by asymmetric dosage regimens (e.g., b.i.d. therapy at 7 a.m. and noon or once-daily controlled release). Although these regimens do not provide 24 hour antianginal and anti-ischaemic effects, they do give a sustained increase in exercise tolerance during the first part of the day with its circadian peak frequencies of anginal attacks (12, 13), and, thus, have become a widely accepted approach (14, 15, 16). So far, little is known about the net effects of these regimens on current health-related quality of life indices. This question is particularly relevant because these compounds produce significant adverse effects such as headache, hypotension, and reflex tachycardia (17).

In 1985, isosorbide mononitrate (ISMN) in a once-daily 30% immediate-release plus 70% sustained-release (IR-SR) formulation was introduced (18). It offered a number of theoretical advantages over conventional isosorbide mononitrate or dinitrate when immediately administered on arising (19).

1. Isosorbide mononitrate provides better bioavailability than isosorbide dinitrate because it had no first-pass metabolism;

2. Once-daily isosorbide mononitrate might be more effective in preventing drug tolerance than multiple-dose regimens given throughout the day;
3. Once-daily instead of multiple-dose therapy might be more convenient for individual patients and might therefore provide better patient compliance;
4. The immediate-release component of the preparation might better protect patients from their circadian peak frequencies of angina pectoris early in the morning.

Beneficial effects of nitrates

The pharmacological effects of the organic nitrates depend on the underlying pathophysiological mechanisms leading to ischaemic heart disease (Table 1). Although early in the development of coronary atherosclerosis the effects of endothelium-dependent vasodilators such as acetylcholine and ATP are usually impaired, the endothelium-independent vasodilator effects of nitrates remain intact (20, 21, 22). Even in saphenous vein bypass grafts this is so (23).

Table 1: Beneficial effects of nitrates.

1.	Vasodilatation of coronary arteries
2.	Reduction of preload and afterload on the heart
3.	Redistribution of flow to vulnerable subendocardial areas
4.	Beneficial effect in hypertensive crisis
5.	Beneficial effect in adult respiratory distress syndrome
6.	Reduction or remodeling
7.	Blockade of platelet aggregation
8.	Blockade of leucocyte adhesion to the vascular endothelium
9.	Protection of endothelial cells from oxidative stress
10.	Antagonism of LDL oxidation

However, with advanced coronary atherosclerosis the development of atheromatous plaques results in narrowing of lumen and restricted blood flow. The peripheral effects of nitrates may become more important. Nitrates produce a vasodilation of the venous vasculature. Dilatation of venous capacitance vessels diminishes venous return to the heart, reducing diastolic volume and pressure. This decreases diastolic wall tension, one of the contributors to the oxygen demand of the heart. Thus, by decreasing preload on the heart, the oxygen needs of the heart diminish. Other effects of nitroglycerin administration also contribute to its beneficial effect in angina. Nitrates cause relaxation of resistance vessels, which decreases afterload placed on the heart. Reducing afterload decreases oxygen demands of the heart, just as reducing preload does. The nitrate effect on resistance arteries generally requires higher concentrations than those needed for venodilatation. Another feature of organic nitrate action in angina pectoris is redistribution of blood flow to the subendocardial areas of the heart, which are especially vulnerable to ischaemia. By decreasing preload, nitrates reduce ventricular filling pressure and increase the time available for endocardial perfusion. In management of angina pectoris caused by coronary artery spasm, the organic nitrates, in addition to effects described above, are useful because of dilatation of constricted coronary vessels. We should add that these beneficial effects of nitrates are not necessarily restricted to angina pectoris, but may occur in any form or expression of heart disease (e.g. different forms of heart failure) (24). Hypertensive crisis and adult respiratory distress syndrome are noncardiac conditions frequently accompanied by heart failure, pulmonary oedema, or angina pectoris and infarction. Nitroglycerin has been unequivocally shown to influence these conditions beneficially and to perform better than nifedipine for that purpose (25, 26, 27).

Newly recognized beneficial effects of nitrates

The pharmacological action of nitrovasodilators appears to be quite similar to that of endothelium derived relaxing factor (EDRF). EDRF is formed in and released from endothelial cells of blood vessels and the heart, and has been shown to be nitric oxide (NO) (28). Vasodilator substances such as acetylcholine, histamine, bradykinin, and adenosine triphosphate act on endothelial cells at their respective receptors to release NO, which subsequently diffuses into the vascular and cardiac smooth muscle cells. It activates cGMP- dependent protein kinase through the second messenger cGMP. This enzyme promotes relaxation of the contractile elements of the cell in several ways, including limiting Ca^{2+} entry through channels and a direct decrease in the sensitivity of contractile proteins to Ca^{2+}. The modification of intracellular calcium levels seems to be the mode of action of the vasoconstrictor compounds norepinephrine, angiotensin II, and serotonin, although a different second messenger (i.e., IP_3) is involved. Recent research has also indicated that these vasoconstrictors stimulate growth factors (i.e., cytokines, vascular endothelial growth factor, insulin, etc.), causing proliferation of fibroblasts, proliferative angiogenesis, vascular muscle cell hypertrophy, apoptosis (programmed cell death) of hypertrophied cells, and atherogenesis. Such effects are currently considered a major mechanism of left ventricular hypertrophy in hypertension and/or remodeling in ischaemic heart disease. Unlike these vasoconstrictors, the vasodilator NO inhibits the proliferation of vascular smooth muscle cells (29, 30) which may contribute to the regression of hypertrophy in such conditions similarly to the regression induced by ACE-inhibitors. However, prospective studies in man are still lacking. Other possibly beneficial effects of NO and its donors have been recognized. NO-donors are not only vasodilators, but also block platelet aggregation, probably by preventing the stimulation of thrombocytic soluble guanylate cyclase (31). Nitrates, in combination with prostacyclin or its analogues, have, therefore, been proposed as a useful antithrombotic therapy (32, 33).

A number of potentially beneficial effects of nitrates have been attributed to their capability to scavenge superoxide. First, they inhibit monocyte and macrophage adhesion to vascular endothelial cells, presumably by antagonizing the oxidation of the lipid cell membrane which triggers the expression of adhesion molecules by endothelial cells in vitro and in vivo (34, 35). Second, they protect the endothelium from oxidative stress, a term frequently used to refer to oxidative damage of DNA and other tissues of the cell leading to cell dysfunction, and ultimately apoptosis (36, 37, 38). Third, they inhibit the 15-lipoxygenase-mediated oxidation of LDL, which is considered an important factor early in the pathogenesis of atherosclerosis (34, 39).

Harmful effects of nitrates

Despite the growing literature attesting to the beneficial effects of NO-donors, other reports have suggested that NO has harmful properties as well. First, it may cause severe hypotension and blood flow abnormalities. Second, a rebound hypertension as well as rebound pulmonary hypertension after withdrawal has been reported (40). Third, it contributes to a catecholamine hyporesponsiveness of the vasculature (41). Fourth, it stimulates negative inotropic cytokines such as TNF, giving rise to myocardial depression (42). Reversible myocardial depression following reperfusion of ischaemic myocardium was documented in patients following myocardial infarction, cardiopulmonary bypass, thrombolytic therapy and coronary angioplasty (43, 44). The condition was associated with increased levels of NO products, and is currently referred to as the stunned myocardium, the mechanism of which was presumed to be cytokine-stimulated and NO-mediated (45). Recent research indicates that rather than NO itself the reaction product of NO with superoxide (O_2^-), peroxynitrite $ONOO^-$, is responsible for the delayed reversibility of cardiac muscle mitochondrial dysfunction, an effect which is prevented by the free radical scavenger urate if present in the blood at physiological levels (36, 46). So, patients with a stunned myocardium may actually benefit from withdrawal of uricosuric compounds, and NO donors may be continued. Fifth, high concentrations of NO have been shown to disrupt metabolism, damage DNA, and cause cell death, suggesting that this free radical might cause direct tissue injury (47). Nitroglycerin intravenously administered in high dosages may give rise to the formation of toxic nitrites which in turn can oxidize the iron of haemoglobin converting it to methaemoglobin. Also, they may give rise to neural toxicity as well as carcinogenity (48).

Clearly, NO does more than just cause vasodilation, and the precise role in humans has not been fully established. Even so, the balance of arguments is currently in its favour. Also with usual dosages (e.g., 5-isosorbide mono- or dinitrate in a dosage of 15–120 mg daily (14)) serious side effects and toxicity are rarely encountered in everyday cardiology practice.

Long term efficacy, side effects and quality of life of isosorbide mononitrate IR-SR formulation.

Efficacy

It is obvious that nitrates can be used safely for long-term treatment. However, little is known about long-term efficacy, side effects and quality of life during continued treatment. Two large, open label studies recently completed addressed these issues (49, 50). A total of 1,350 patients with stable angina pectoris were treated for three months with conventional ISMN/ISDN 1,020 mg twice or three times daily, and subsequently, for the same period with once-daily 50 mg IR-SR formulation (49). A total of 1,212 patients could be evaluated after six months. Not only did patients experience a 22.2% ($p = 0.03$) reduction of anginal attack frequency during the second period, but also drug compliance improved by 23.8% ($p = 0.02$). The investigators subsequently tested the hypothesis that 100 mg IR-SR formulation may perform even better than 50 mg; they used the same study design for assessment. In 505 patients with stable angina pectoris, 100 mg IR-SR for three months produced a further 13.9% ($p = 0.001$) and 2% ($p = 0.01$) improvement of patient compliance. The better compliance may explain part of the improved efficacy, but not all of it. Table 2 shows a comparison of multiple dose and 100 mg IR-SR formulation as calculated by the product of odds ratios of "multiple dose/50 mg IR-SR" and "50 mg IR-SR/100 mg IR-SR". Although this is an extrapolation, it may be acceptably valid since the procedure is straightforward and the data are from similar institutions and investigators (19). The largely unchanged quantity of sublingual nitroglycerin required in the studies was consistent with the situation in which an improved angina classification is associated with increased physical activities.

Side effects

Drug tolerance did not develop after 50-100 mg IR-SR formulation, patients' self-assessed rating scores continued to improve rather than to deteriorate on higher dosages (49, 50). Drug tolerance did not develop in three more open label trials from other groups ($n = 30$–45 patients per study) (51, 52, 53) after 6–13 months of treatment, whereas improvement of anginal attack frequencies, and level of ST-depression during angina were similar to those in the above two studies. Except for headache, other side effects of 50-100 mg isosorbide mononitrate IR-SR formulation were rare and negligible. Headache, nonetheless, occurred as frequently as with multiple ISMN/ISDN dosages in between 41 and 67% of the patients (49, 50). Particularly during the first weeks of treatment, headache is commonly reported. For example, in a six-week double-blind, parallel-group study, both IR-SR formulation 50–100 mg once daily and conventional isosorbide mononitrate 80–120 mg

Table 2: Odds ratios* of symptoms and Quality of life indices of patients with stable angina pectoris treated with 50 mg and 100 mg isosorbide mononitrate IR-SR formulation once daily vs. standard multiple dose ISMN/ISDN.

	odds ratios* multiple dose/50 mg IR-SR	p value	odds ratios* multiple dose/100 mg IR-SR	p value
mobility index	0.83	< 0.001	0.69	< 0.001
side effect index	0.99	0.85	0.96	< 0.001
anginal pain index	0.64	< 0.001	0.56	< 0.001
anginal pain index 8–10 hr am	0.65	0.006		
psychological well-being index	0.87	0.036	0.75	< 0.001
patient compliance	0.50	< 0.001	0.48	< 0.001
sublingual nitroglycerin consumption	0.94	0.68		

*odds ratios = mean scores during multiple-dose conventional isosorbide mononitrate/mean score during IR-SR formulation.

daily necessitated drug withdrawal in up to 17% of the patients (54). In a double-blind study by our group, (55) it was demonstrated that this effect can be avoided to a great extent by starting with isosorbide mononitrate dosages as low as 25-30 mg daily. Such dosages of IR-SR formulation have become available recently.

Quality of life

Important quality of life indices were also addressed in two studies (49, 50). A once-daily dose of 50 mg IR-SR for three months gave significantly better scores than multiple-dose conventional ISMN/ISDN: mobility index improved 5.5% ($p = 0.04$), and psychological well-being index improved 5.6% ($p = 0.04$). In addition, 100 mg IR-SR once daily for three months gave better scores than 50 mg IR-SR: 20.3% ($p < 0.001$) and 5.8% ($p < 0.001$) further improvement of mobility and psychological well-being indices. Table 2 shows the extrapolated odds ratios of multiple dose/100 mg IR-SR calculated by the products of odds ratios "multiple dose/50 mg IR-SR" and "50 mg IR-SR/100 mg IR-SR" from the two separate studies (49, 50). Although based on extrapolation, these estimates suggest that in particular high-dose IR-SR formulation may provide a considerable further improvement of mobility and psychological well being.

Discussion

Unfortunately, none of the long term studies on IR-SR isosorbide mononitrate formulation were blinded, and so the risk of placebo effects partly explaining the results cannot fully be ruled out. A current problem with designing long-term nitrate studies is that it is ethically increasingly difficult to replace nitrate therapy with placebo in symptomatic patients even though improved survival with long-term nitrate therapy has not been unequivocally proven. At the least, the results of GISSI-3 and ISIS-4 (5, 6) do not support the presence of major placebo effects in the data, since important changes occurred in the mobility and psychological well-being indices rather than in the anginal pain and side effect indices, which are generally particularly susceptible to placebo effects in QOL assessments (2).

The primary approach of patients with angina pectoris must involve the adequate treatment of their coronary condition, e.g. by coronary bypass surgery or percutaneous coronary angioplasty, rather than treatment with nitrates. If patients remain symptomatic despite adequate treatment, or if they have to wait for such interventions or do not qualify for them for any other reason, nitrates as well as other anti-anginal drugs should be considered. Parker and Parker (56) recently reviewed prescribing patterns regarding current nitrate therapy and recommended that (1) slow-release nitrates, rather than beta-blockers or calcium channel blockers, be given as an initial preventive therapy for stable angina pectoris, particularly in patients who respond well to sublingual nitroglycerin; and (2) beta-blockers or calcium channel blockers be given as initial therapy in patients with coexistent hypertension and/or a history of MI. We suggest that nitrates should be routinely added in the latter category, given their apparent improvement of various QOL indices. A particularly interesting aspect of treatment with nitrates, although predominantly of a theoretical nature so far, is their potential to offer more than just symptomatic treatment (17). Nitrates, being NO donors, have been thought to reduce remodelling (29), block platelet aggregation (30), and adhesion receptors of vascular endothelial cells (35), and probably most importantly to protect the endothelium from oxidative stress by scavenging superoxide (38). Angiotensin II is an important producer of oxygen free radicals and it is this property that is largely held responsible for the deleterious effects of the compound on the myocardium as well as vascular endothelium (57). A supposed antagonistic effect between NO and angiotensin II on oxygen free radicals is a plausible hypothesis (58) to explain the widely published beneficial effect of ACE inhibitors on the efficacy of nitrate treatment (59). The risk of hypotension with nitrates is small because the compound has little influence on systemic blood pressure and the same applies with the IR-SR formulation. The risk of methaemoglobin due to overdosing (14) should be better prevented by a once-daily IR-SR formulation with predictable pharmacokinetic pattern than by multiple dosages of conventional nitrates.

Conclusion

We conclude that IR-SR isosorbide mononitrate has a rapid onset of action, is clinically efficient, and provides better quality of life than conventional ISMN/ISDN. Tolerance with continued use of the formulation has not yet been reported. Long-term efficacy data both of isosorbide mononitrate IR-SR and conventional ISMN/ISDN are limited so far. Large studies in patients with angina pectoris (e.g. International Quality Of Life assessment of patients with Angina pectoris on Nitrate therapy – IQOLAN) and in patients with heart failure (e.g. DUtch Mononitrate Quality of Life-Heart Failure – DUMQOL-HF) addressing long term effects, including morbidity and mortality data, are ongoing and some of the data should be completed within the next few months.

Acknowledgements

We wish to thank Guido de Regt, from BYK Medical Department, Zwanenberg, The Netherlands for his support on this manuscript.

References

1. Mayou R, Bryant B. Quality of life in cardiovascular disease. Br Heart J 1993; 69: 460–466.
2. Thompson D, Meadows K, Lewin R. Measuring quality of life in patients with coronary heart disease. Eur Heart J 1998; 19: 693–695.
3. Morris J, Cowan J. Nitrates in myocardial infarction: a current perspective. Can J Cardiol 1995 11: 5B–10B.
4. Jugdutt B. Nitrates in myocardial infarction: a review. Cardiovasc Drugs Ther 1994; 8: 635–646.
5. Gruppo Italiano per lo Studio della Soppravivenza nell' infarto Miocardico (GISSI-3). Effects of lisinopril and transdermal glyceryl trinitrate singly and together on 6-week mortality and ventricular function after acute myocardial infarction. Lancet 1994; 343: 1115–1122.
6. ISIS-4 collaborative group. A randomised factorial trial assessing only oral captopril, oral mononitrate, and intravenous magnesium sulphate in 58,050 patients with suspected acute myocardial infarction. Fourth International Study of Infarct Survival. Lancet 1995; 345: 669–685.
7. Kugiyama K, Yasue H, Okumara K. Nitric oxide activity is deficient in spasm arteries of patients with coronary spastic angina. Circulation 1996; 94: 266–272.
8. Diver D, Ferreira P, Sharaf B, et al. Clinical and arteriographic characterization of patients with unstable angina without critical coronary arterial narrowing. Am J Cardiol 1994; 74: 531–537.
9. Nijberg G, Carleus P, Lindstrom E. The effect of isosorbide-5-mononitrate Durules on exercise tolerance in patients with exertional angina pectoris. A placebo controlled study. Eur Heart J 1986; 7: 835–842.
10. Parker J, Wisenberg G. Antianginal effects of sustained release isosorbide-5-mononitrate. Circulation 1989; 80/s2: 267.
11. Watanabe H, Kakihana M, Ohtsuka S, et al. Platelet cyclic GMP, a potentially useful indicator to evaluate the effects of nitroglycerin and nitrate tolerance. Circulation 1993; 88: 29–36.

12. Scheidel B, Lenhard G, Blume H. Chronopharmacology of isosorbide-5-mononitrate (immediate release, retard formulation). Eur J Clin Pharmacol 1989; 36: 177–178.
13. Lemmer D, Scheidel B, Blume H, et al. Clinical chronopharmacology of oral sustained-release isosorbide-5-mononitrate. Eur J Clin Pharmacol 1991; 40: 71–75.
14. Thadani U, Lipicky R. Short and long-acting oral nitrates for stable angina pectoris. Cardiovasc Drugs Ther 1994; 8: 611–623.
15. Nordlander R. Can nitrate tolerance be avoided with once-daily administration of isosorbide-5-mononitrate 60 mg in Durules? Eur Heart J 1989; 10/s: 73.
16. Klemsdal U, Giesdal R. Intermittent or continuous transdermal nitroglycerin: still an issue, or is the case closed? Cardiovasc Drugs Ther 1996; 10: 5–11.
17. Niemeyer M, Cleophas T. Nitrates may provide benefits other than just symptomatic treatments. Cardiologie 1997; 4: 205–208.
18. Igendaay A, Klein G, Rehm K. Vergleichende Pharmacokinetik retardierter Mononitratpräparate. Münch Med Wochenschr 1987; 129: 34–36.
19. Cleophas T, Niemeyer M, Zwinderman A, et al. A new immediate/sustained-release formulation of isosorbide mononitrate. Cardiologie 1999; 6: 358–363.
20. Conti C. Why use nitrate in 1990? Eur Heart J 1991; 12 (suppl. A.): 2–4.
21. Anggard E. Endogenous nitrates-implications for treatment and prevention. Eur Heart J 1991; 12 (suppl. A): 5–8.
22. Parker J. Nitrates and angina pectoris. Am J Cardiol 1993; 72: 3C–8C.
23. Berglund H, Luo H, Nishioka T, et al. Preserved vasodilatory response to nitroglycerin in saphenous vein bypass grafts. Circulation 1996; 94: 2871–2876.
24. Remme W. Pathophysiology and therapy of heart failure, New insights and developments. Part ll Neth J Med 1993; 43: 125–146.
25. Bussmann W, Kenedi P, von Mengelen H, et al. Comparison of nitroglycerin with nifedipine in patients with hypertensive crisis. Kardiologie 1993; 82: 33–37.
26. Rossiant R, Falke K, Lopez F, et al. Nitric oxide for adult respiratory distress syndrome. N Engl J Med 1993; 328: 299–305.
27. Kaplan N. Treatment of hypertension. In: Clinical Hypertension (ed. NM Kaplan), Wilkens & Wilkens, Baltimore, 1994, 145–170.
28. Ignarro LJ, Byrns R, Wood K. Pharmacological and biochemical properties of endothelium-derived relaxing factor (EDRF): Evidence that EDRF is closely related to nitric (NO) radical. Circulation 1986; 74: II–287.
29. Garg V, Hassid A. Nitric oxide-generating vasodilators and cyclic guanosine monophosphate inhibit mitogenesis of rat vascular smooth muscle cells. J Clin Invest 1989; 83: 1774–1777.
30. Johnson D, Dell Italia L. Cardiac hypertrophy and failure in hypertension. Current Opin Nephrol Hypertens 1996; 5: 186–191.
31. Graaf de J, Bange J, Moncada S, et al. Nitric oxide functions as an inhibitor of platelet adhesion under flow conditions. Circulation 1992; 85: 2284–2290.
32. Radomski M, Moncada S. Biological role of nitric oxide in platelet function. In: Clinical Relevance of Nitric Oxide in the Cardiovascular System (eds. S Moncada, E Higgs, J Berrazueta), Edicomplet, Madrid, 1991, 45–56.
33. Sinzinger H, Rauscha F, O'Grady J, et al. Prostaglandin I2 and the nitric oxide donor molsidomine have synergistic effects on thromboresistance in man. Br J Clin Pharmacol 1992; 33: 289–292.
34. Catapano A. Antioxidant effect of flavonoid and nitric oxide. Angiology 1997; 48: 39–44.
35. Kubes P, Suzuki M, Granger D. Nitric oxide: an endogenous modulator of leukocyte adherence. Proc Natl Acad Sci USA 1991; 88: 4651–4655.
36. Rongen G, Smits P, Thien T. Endothelium and the role of nitric oxide. Neth J Med 1994; 44: 26–35.

37. Sarafin T, Bredsen DE. Is apoptosis mediated by reactive oxygen species? Free Radic Res 1994; 20: 1–6.
38. Stamler J. Redox Signaling: nitrosylation and related target interactions of nitric oxide. Cell 1994; 78: 931–936.
39. Steinberg D, Partasarathy S, Carew T, et al. Beyond cholesterol, modifications of low-density lipoprotein that increase its atherogenicity. N Engl J Med 1989; 320: 915–924.
40. Warren J, Higgenbottam T. Caution with use of nitric oxide. Lancet 1996; 348: 629–630.
41. Finkel M, Oddis C, Jacob T, et al. Negative inotropic effects of cytokines on the heart mediated by nitric oxide. Science 1992; 257: 387–389.
42. Nathan C, Xie Q: Nitric oxide synthases: roles, tolls, controls. Cell 1994; 78: 915–918.
43. Braunwald E, Kloner R. The stunned myocardium: prolonged, postischemic ventricular dysfunction. Circulation 1982; 66: 1146–1149.
44. Dilsizian V, Bonow R. Current diagnostic techniques of assessing myocardial viability in patients with hibernating and stunned myocardium. Circulation 1993; 86: 1–20.
45. Finkel M, Oddis C, Hattler B, et al. Myocardial ischemia, stunning and hibernation. In : Myocardial Viability (eds. AS Iskandrian, EE van der Wall), Kluwer Academic Publishers, Dordrecht, 1994, 5–18.
46. Xie Y, Wolin M. Role of nitric oxide and its interaction with superoxide in the suppression of cardiac muscle mitochondrial respiration, involvement in response to hypoxia/reoxygenation. Circulation 1996; 94: 2580–2586.
47. Nathan C. Nitric oxide as a secretory product of mammalian cells. FASEB J 1992; 6: 3051–3064.
48. Moncada S, Higgs A. The L-arginine-nitric oxide pathway. N Engl J Med 1993; 329: 2002–2012.
49. Niemeyer M, Kleinjans H, De Ree R, et al. Comparison of multiple-dose and once-daily nitrate therapy in 1,212 patients with stable angina pectoris: effects on quality-of-life indices. Angiology 1997; 48: 855–863.
50. Zwinderman A, Cleophas T, van der Sluis H, et al. Effects of 50 mg and 100 mg isosorbide mononitrate once daily on quality of life in 505 patients with stable angina pectoris. Angiology (in press).
51. Krepp H. Langzeitbehandlung der koronaren Herzerkrankung mit elantan®long. In: Mononitrat IV (eds. U Borchard, W Rafflenbeul, A Schrey, A.), Verlag Wolf & Sohn, Munich 1985, 176–183.
52. Heepe W, Gathmann-Lewik U. Antianginal efficacy and tolerability of 50 mg sustained-release isosorbide mononitrate in an open twelve-month observation study. Cardiology 1987; 74: 34–39.
53. Ahmadinejad M, Eghbal B, Sorgenicht W, et al. Slow-release isosorbide mononitrate: a new once daily therapeutic modility for angina pectoris. Eur Heart J 1988; 9: 135–139.
54. Walker J, Curry P, Bailey A. A Comparison of nifedipine once daily, isosorbide mononitrate once daily, and isosorbide dinitrate twice daily in patients with chronic stable angina. Int J Cardiol 1996; 79: 117–126.
55. Cleophas T, Niemeyer M, van der Wall E. Nitrate-induced headache in patients with stable angina pectoris: beneficial effect of starting on a low dosage. Am J Ther 1996; 3: 802–806.
56. Parker J, Parker J. Nitrate therapy for stable angina pectoris. N Engl J Med 1998; 338: 520–553.
57. Cleophas T, van der Meulen J, Niemeyer M, et al. Mechanisms offsetting the beneficial effects of antihypertensive drugs. Perfusion 1998; 11: 373–380.
58. Münzel T, Sayegh H, Freeman B, et al. Evidence for enhanced vascular superoxide anion production in nitrate tolerance. A novel mechanism underlying tolerance and cross-tolerance. J Clin Invest 1995; 95: 187–194.
59. Münzel T, Bassenge E. Long-term angiotensin-converting inhibition with high-dose enalapril retards nitrate tolerance in large epicardial arteries and prevents rebound coronary vasoconstriction in vivo. Circulation 1996; 93: 2052–2058.

New trends in interventional cardiology

P.G. Steg

Introduction

Since its inception in 1977, percutaneous coronary intervention has bloomed into the main technique for revascularization of patients with coronary artery disease. Refinements in the technique, the equipment, as well as in adjuvant therapy, have yielded tremendous progress in efficacy and safety. Interventional cardiology has become the mainstay of therapy for stable and unstable forms of coronary artery disease, including acute myocardial infarction. Recently, the advent of intracoronary stents and of very effective antiplatelet therapy using thienopyridines and glycoprotein IIb/IIIa receptor antagonists have set new standards for performance.

Future avenues for research and improvement include:

- new therapies to prevent or cure restenosis: restenosis after successful coronary intervention remains a lingering and disturbing problem, leading to recurrent revascularization in up to 20% of patients. Even though intracoronary stents have reduced its incidence, restenosis has not disappeared and in-stent restenosis has a high propensity for recurrence. Current areas of clinical investigation include new pharmacological or even gene therapy to prevent constrictive remodeling, intimal hyperplasia or inflammation, or to improve healing at the treated site. Alternatively, intracoronary irradiation using beta or gamma emitters (brachytherapy) has also had some clinical success.
- therapeutic angiogenesis using vascular growth factors, gene therapy or even laser-induced myocardial revascularization.
- new physiological tools for the selection of patients and assessment of results of intervention (e.g. flow wire assessment of intracoronary velocities, pressure wire assessment of pressure gradients, etc.)
- outcome analysis of the benefit of percutaneous intervention, as compared to alternative treatments, both in stable and unstable forms of the disease, to assess the optimal indications and determine health policies.

There has been both tremendous growth and tremendous progress in the field of interventional cardiology in the past ten years. Percutaneous coronary intervention has become the major method for myocardial revascularization and the volume of coronary intervention has grown steadily in all industrialized countries. It is estimated that in 1995, more than 200,000 percutaneous transluminal coronary angioplasties were performed in Europe and, despite marked variation between countries in terms of volume, the trend curves indicate consistent steady and continuing growth, already reaching a level of more than 1,000 interventions/million inhabitants in the countries with the highest volumes per capita (Germany and Belgium). This article will review briefly some of the most important advances of the recent years, as well as outline some of the avenues open for continuing improvement.

Coronary stents

Perhaps the single most important advance in interventional cardiology in the 90s has been the advent of intracoronary stents. While the first intracoronary stents were placed in man as early as 1984, it was only in 1992-94 that stent use started growing, largely due to two factors: first, the clear demonstration by randomized clinical trials, such as BENESTENT and STRESS, of the short- and long-term benefits of stents over conventional balloon angioplasty (1, 2), with both increased success rates and decreased procedural complications, as well as reduced restenosis rates. Second, the demonstration by registries and then by randomized clinical trials (3), that a simple oral regimen of combined antiplatelet therapy, using aspirin and ticlopidin was both safe and extremely effective at preventing coronary stent thrombosis, avoiding the need for prolonged anticoagulation, or the use of intravenous heparin and oral anticoagulants. This helped to reduce the complication rates, hospital stay, and ultimately costs associated with stenting. The ultimate effect of these trials was an exponential growth in the use of stents in the catheterization laboratories, culminating in a true "stentomania" in some countries, such as France, where the rate of stenting exceeds 75% of the procedures. This was also associated with an increase in the diversity of stent designs available: while most of the early studies and clinical experience had been performed using the articulated Palmaz-Schatz PS 153 stent in discrete (15 mm length) lesions in relatively large (3 mm) coronary arteries, later experience has involved more than 50 different stent designs that are currently being used in a wide variety of target lesion types. One of the important consequences of the advent of stents is that it has nearly abolished the need for surgical standby. The rate of emergency coronary artery bypass grafting has dwindled to less than 1% in all centers, thanks to the ability of stents to "seal" the arterial dissections generated by balloon angioplasty, which constituted the substrate for acute vessel closure and its attendant potentially catastrophic consequences.

Improved antiplatelet therapy

Traditionally, interventional procedures are performed under pre-treatment with aspirin and with intravenous heparin. However, aspirin is an imperfect antiplatelet agent. Glycoprotein IIb/IIIa inhibitors and specifically the chimeric monoclonal antibody against GpIIb/IIIa, abciximab, have been extensively tested as adjuncts to coronary interventional procedures with the hope of reducing peri-procedural complications and ultimately improving clinical outcome. The results of the many trials performed in the catheterization laboratory with abciximab have been remarkably consistent in demonstrating substantial clinical benefit of this treatment. The EPIC trial first demonstrated improved outcome in high-risk patients undergoing PTCA for unstable angina or acute myocardial infarction, receiving a combination of bolus dose and follow-up infusion for 12 hours of abciximab. This was subsequently confirmed by the CAPTURE trial, in which patients with refractory unstable angina undergoing abciximab treatment prior to angioplasty had markedly improved outcome. Then, the EPILOG study showed that the benefit of abciximab could be extended to all patients undergoing elective percutaneous coronary interventions and that the bleeding risks associated with such potent antiplatelet therapy could be nearly abolished by reducing the dose of heparin and by meticulous care of the arterial access site. Finally, the EPISTENT study showed that the benefit observed in previous studies could be extended to the stent era, and that in fact the combination of stenting and abciximab was optimal in reducing both short and long-term complications of percutaneous coronary intervention, setting a new standard for the results of interventional cardiology (4). As in the case of stents, these consistent clinical trial results have led to marked increases in the rate of use of such effective antiplatelet therapy, albeit still somewhat limited by the relatively high price of this agent, and to the advent of a variety of other agents able to block the GpIIb/IIIa receptor via other mechanisms and which have also shown benefit in large clinical trials (PURSUIT, PRISM, PRISM+), mostly in patients with unstable angina. The only disappointment with glycoprotein IIb/IIIa antagonists is that they appear to have no impact on restenosis, despite the fact that the EPIC trial follow-up found improved long-term outcome in patients treated initially with the bolus and infusion of abciximab (5).

Restenosis: a lingering problem

While stents have markedly reduced the rate of recurrence after successful coronary angioplasty, restenosis has not been abolished and still occurs in 4–20% of patients, varying with the patient population, the methods used in clinical trials, and also importantly with the definitions used. Even worse is the fact that despite the overall beneficial effect of stents, in-stent restenosis, when it occurs, appears to have a high propensity for recurrence, leading to repeated revascularizations. In addition, in recent years we have witnessed a change in the pathophysiological conception of restenosis: while it was classically viewed as a simple phenomenon of intimal thickening due to smooth muscle cell proliferation and extracellular matrix production in response to the stretch and injury produced by balloon angioplasty, consistent experimental and clinical observations (in particular using intravascular ultrasound) made during the early 90s have demonstrated that restenosis after balloon angioplasty was mostly the consequence of a phenomenon of constrictive remodeling, whereby the entire cross-sectional area of the vessel is narrowed (6–9), although the exact determinants of remodeling remain unclear (10). In fact, this appears to be one of the reasons why stents are so effective in preventing restenosis: by providing a stable, incompressible metallic scaffold to the lumen, they nearly abolish constrictive remodeling. However, it was subsequently found that there was a phenomenon of reduced luminal area at stented sites, due to intimal thickening in the struts region. There is increasing evidence that this is at least in part related to inflammation. There are therefore several methods available to attempt to reduce restenosis.

Pharmacological therapy

Probucol, a lipid lowering drug with antioxidant properties, has demonstrated clinical efficacy in preventing restenosis (11, 12), but only in patients undergoing balloon angioplasty (as opposed to stenting) and requires pre-treatment for several weeks prior to PTCA.

Tranilast, an anti-TGFß drug, has shown experimental and clinical promise (13–15) and is undergoing large-scale clinical testing.

IL-10, an anti-inflammatory cytokine, has been shown in our laboratory to inhibit macrophage activation and infiltration and to prevent intimal thickening and preserve the arterial lumen in experimental models of restenosis following balloon angioplasty as well as arterial stenting (16).

Brachytherapy

This designates the use of intracoronary radiation, with the hope of reducing the restenosis rate. Various devices are available to deliver radiation, using ß- or γ-emitters, and ranging from a radiation wire temporarily inserted in the coronary artery during or at the end of the angioplasty procedure, to permanent radioactive stents (17). Several large clinical trials are currently ongoing to test the validity of this strategy. Small-scale clinical studies have shown the feasibility of this strategy (18) and the efficacy of radiation therapy in reducing the restenosis rate in patients at high risk (19, 20). Obviously, the concerns with this type of approach would be the fear of acute adverse events related to irradiation, such as intracoronary thrombosis, and most of all the fear of late fibrosis or aneurysm formation at the irradiated site. This is a legitimate concern due to the difficulties in determining the appropriate dosimetry, given the fact that most (approximately 70 %) intracoronary atherosclerotic plaques are eccentric. Confirmatory data and a fairly long follow-up are required before the clinical safety of these devices can be ensured with confidence (21, 22).

Gene therapy

It is based upon the hypothesis that it may be possible immediately after (or even during) PTCA to deliver via a catheter a gene to arterial wall cells, and that the expression of the gene, even if transient, may inhibit migration and/or proliferation of medial smooth muscle cells at the dilated site or extracellular matrix production (23, 24). Due to the lack of knowledge regarding the precise mechanism of constrictive remodeling, prevention of the latter appears for now to rest mainly on mechanical methods such as stenting rather than on gene therapy. Effective gene delivery systems are now available for arterial gene transfer. Macro and micro-transfer should be distinguished. Micro-transfer is the use of a vector (viral vector, plasmid-containing liposome, or even "naked" DNA) able to transfer a gene to target cells in the arterial wall, i.e. mostly to medial smooth muscle cells. "Macrotransfer" corresponds to the local delivery system used to bring the vector in contact with the arterial wall. It is often a catheter (25) or a stent. The most efficient vectors in terms of transfer efficiency are currently recombinant replication-defective adenoviral vectors. Several groups, including ours, have demonstrated that percutaneous local catheter delivery of adenoviral vectors allowed efficient transfer to arterial wall cells and in particular to the endothelium (26). However, despite the design of effective tools for percutaneous gene transfer, several important problems remain:

- Transgene expression is transient (after adenoviral vector-mediated transfer) and vanishes after several weeks, most likely because of a powerful and early inflammatory and immune response (both cellular and humoral), triggered by the expression of adenoviral proteins. Nevertheless, for prevention of restenosis, in itself a transient phenomenon occurring within weeks of the initial arterial trauma, transient expression of the transgene is not incompatible with therapeutic efficacy.
- Transfer efficiency is high in normal arteries, but decreases markedly when the target vessels are atheromatous, which implies the use of adjuvant therapies likely to substantially increase transfer efficiency or the use of therapeutic strategies relying on transfer of a gene whose product is secreted and therefore liable to exert therapeutic effects on nontransduced cells (27).
- There is a risk of generation of replication-competent recombinant viral vectors.

It is therefore imperative to design safer vectors, which may limit or avoid the immune response, limit the risk of generating replication-competent viruses, and may even be targeted toward specific tissues, such as arterial smooth muscle, by using tissue-specific promoters. Despite these limitations, gene therapy has proved effective in the prevention of experimental restenosis in models of intimal hyperplasia induced by arterial trauma (28). A host of therapeutic genes have been tested experimentally.

Cytotoxic strategies have used genes that encode proteins that will ultimately kill proliferating cells. These strategies can be refined by making cytotoxicity selective for proliferating cells: an example is the thymidine kinase gene of HSV-1, which will kill proliferating smooth muscle cells (by inducing apoptosis) if these receive a sensitizing agent (ganciclovir or acyclovir). This strategy was proven effective in animal models of restenosis in normal and atheromatous vessels (29–31).

Cytostatic strategies aim at stopping division of the transfected cells without killing the cells. In contrast to cytotoxic strategies, this would prevent the release of inflammation mediators and mitogenic factors by the cell, which might in turn lead to activation of neighboring cells or tissue lesions. One shortcoming is that these strategies rely on genes encoding intracellular proteins. Therefore, to stop intimal hyperplasia, one has to use very effective transduction vectors in order to stop the cell cycle in a sufficient number of target cells. Transfer of the Rb gene, cell cycle protein inhibitors (32, 33), or of the gax gene (34), a negative regulator of proliferation specific for vascular smooth muscle cells, are typical examples of this type of strategy.

Other strategies aim at speeding regeneration of a normal endothelium (35), which appears to inhibit the development of intimal hyperplasia in the underlying vessel layers, using transfer of the VEGF gene for instance (36), or at mimicking some endothelial functions by transferring NO synthases (37–40).

The time is ripe for clinical trials of these various strategies that have shown consistent benefit in a host of experimental models. If an effective cure could be found for the lingering problem of restenosis, there would be "no limit" to the future of interventional cardiology.

References

1. Fischman D, Leon M, Baim D, et al. A randomized comparison of coronary-stent placement and balloon angioplasty in the treatment of coronary artery disease. N Engl J Med 1994; 331: 496–501.
2. Serruys PW, de Jaegere P, Kiemeneij F, et al. A comparison of balloon-expandable-stent implantation with balloon angioplasty in patients with coronary artery disease. N Engl J Med (1994; 331: 489–495.
3. Schömig A, Neumann FJ, Kastrati A, et al. A randomized comparison of antiplatelet and anticoagulant therapy after the placement of coronary artery stents. N Engl J Med 1996; 334: 1084–1089.
4. Topol E, Serruys P. Frontiers in interventional cardiology. Circulation 1998; 98: 1802–1820.
5. Topol EJ: Randomized trial of coronary intervention with antibody against platelet IIb/IIIa integrin for reduction of clinical restenosis: results at 6 months. Lancet 1994; 343: 881–886.
6. Lafont A, Guzman, et al. Restenosis after experimental angioplasty. Intimal, medial, and adventitial changes associated with constrictive remodeling. Circ Res 1995; 76: 996–1002.
7. Kakuta T, Currier JW, Haudenschild CC, et al. Differences in compensatory vessel enlargement, not intimal formation, account for restenosis after angiolasty in the hypercholesterolemic rabbit model. Circulation 1994; 89: 2809–2815.
8. Mintz GS, Popma JJ, Pichard AD, et al. Arterial remodeling after coronary angioplasty. A serial intravascular ultrasound study. Circulation 1996; 94: 35–43.
9. Mintz GS, Pichard AD, Kent KM, et al. Endovascular stents reduce restenosis by eliminating geometric arterial remodeling: a serial intravascular ultrasound study. J Am Coll Cardiol 1995; 25: 36A (Abstract).

10. Le Feuvre C, Tahlil O, Paterlini P, et al. Arterial response to mild balloon injury in the normal rabbit: evidence for low proliferation rate in the adventitia. Coron Artery Dis 1998; 9: 805–814.
11. Cote G, Tardif JC, Lesperance J, et al. Effects of probucol on vascular remodeling after coronary angioplasty. Multivitamins and Protocol Study Group. Circulation 1999; 99: 30–35.
12. Tardif JC, Cote G, Lesperance J, et al. Probucol and multivitamins in the prevention of restenosis after coronary angioplasty. Multivitamins and Probucol Study Group (see comments). N Engl J Med 1997; 337: 365–372.
13. Takahashi A, Taniguchi T, Ishikawa Y, et al. Tranilast inhibits vascular smooth muscle cell growth and intimal hyperplasia by induction of p21(wafl/cipl/sdil) and p53. Circ Res 1999; 84: 543–550.
14. Kosuga K, Tamai H, Ueda K, et al. Effectiveness of tranilast on restenosis after directional coronary atherectomy. Am Heart J 1997; 134: 712–718.
15. Fukuyama J, Ichikawa K, Hamano S, et al. Tranilast suppresses the vascular intimal hyperplasia after balloon injury in rabbits fed on a high-cholesterol diet. Eur J Pharmacol 1996; 318 : 327–332.
16. Feldman L, Aguirre L, Ziol M, et al. Interleukin-10 inhibits intimal hyperplasia after angioplasty or stent implantation in hypercholesterolemic rabbits. Circulation (1999) in press.
17. Fischell TA. Radioactive stents. Semin Interv Cardiol 1998; 3: 51–56.
18. King SB, Williams DO, Chougule P, et al. Endovascular beta-radiation to reduce restenosis after coronary balloon angioplasty: results of the beta energy restenosis trial (BERT). Circulation 1998; 97: 2025–2030.
19. Condado JA, Waksman R, Gurdiel O, et al. Long-term angiographic and clinical outcome after percutaneous transluminal coronary angioplasty and intracoronary radiation therapy in humans. Circulation 1997; 96: 727–732.
20. Teirstein PS, Massullo V, Jani S, et al. Two-year follow-up after catheter-based radiotherapy to inhibit coronary restenosis (see comments). Circulation 1999; 99: 243–247.
21. King SB III: Radiation for restenosis: watchful waiting. Circulation 1999; 99: 192–194.
22. Serruys PW, Levendag PC. Intracoronary brachytherapy: the death knell of restenosis or just another episode of a never-ending story? Circulation 1997; 96: 709–712.
23. Steg PG, Feldman LJ: Prospects for gene therapy of postangioplasty restenosis. Adv Nephrol Necker Hosp 1997; 26: 121–131.
24. Steg PG, Feldman L: Prospects for gene therapy of post-angioplasty restenosis. In: Seminars in Nephrology (eds. J Grünfeld, J Bach, H Kreis), MH.Maxwell Ed. Mosby-Year book, Chicago, 1996, 133–144.
25. Tahlil O, Brami M, Feldman LJ, et al. The Dispatch catheter as a delivery tool for arterial gene transfer. Cardiovasc Res 1997; 33: 181–187.
26. Varenne O, Gerard RD, Sinnaeve P, et al. Percutaneous adenoviral gene transfer into porcine coronary arteries: is catheter-based gene delivery adapted to coronary circulation? Hum Gene Ther 1999; 10: 1105–1115.
27. Feldman LJ, Pastore CJ, Aubailly N, et al. Improved efficiency of arterial gene transfer by use of poloxamer 407 as a vehicle for adenoviral vectors. Gene Ther 1997; 10: 189–198.
28. Feldman LJ, Steg G. Optimal techniques for arterial gene transfer. Cardiovasc Res 1997; 35: 391–404.
29. Steg PG, Tahlil O, Aubailly N, et al. Reduction of restenosis after angioplasty in an atheromatous rabbit model by suicide gene therapy. Circulation 1997; 96: 408–411.
30. Ohno T, Gordon D, San H, et al. Gene therapy for vascular smooth muscle cell proliferation after arterial injury. Science 1994; 265: 781–784.
31. Simari RD, San H, Rekhter M, et al. Regulation of cellular proliferation and intimal formation following balloon injury in atherosclerotic rabbit arteries. J Clin Invest 1996; 98: 225–235.
32. Chang MW BE, Seltzer J, Jiang Y-Q, et al. Cytostatic gene therapy for vascular proliferative disorders with a constitutively active form of the retinoblastoma gene product. Science 1995; 267: 518–522.

33. Chang MW, Barr E, Lu, et al. Adenovirus-mediated over-expression of the cyclin/cyclin-dependent kinase inhibitor, p21 inhibits vascular smooth muscle cell proliferation and neointima formation in the rat carotid artery model of balloon angioplasty. J Clin Invest 1995; 96: 2260–2268.
34. Maillard L, Van Belle E, Smith RC, et al. Percutaneous delivery of the gax gene inhibits vessel stenosis in a rabbit model of balloon angioplasty. Cardiovasc Res 1997; 35: 536–546.
35. Asahara T, Chen D, Tsurumi Y, et al. Accelerated restitution of endothelial integrity and endothelium-dependent function after phVEGF165 gene transfer. Circulation 1996; 94: 3291–3302.
36. Van Belle E, Tio FO, Chen D, et al. Passivation of metallic stents after arterial gene transfer of phVEGF165 inhibits thrombus formation and intimal thickening. J Am Coll Cardiol 1997; 29: 1371–1379.
37. Carmeliet P, Moons L, Dewerchin M, et al. Insights in vessel development and vascular disorders using targeted inactivation and transfer of vascular endothelial growth factor, the tissue factor receptor, and the plasminogen system. Ann N Y Acad Sci 1997; 811: 191–206.
38. von der Leyen HE, Gibbons GH, Morishita R, et al. Gene therapy inhibiting neointimal vascular lesion: in vivo transfer of endothelial cell nitric oxide synthase gene. Proc Natl Acad Sci USA 1995; 92: 1137–1141.
39. Varenne O, Pislaru S, Gillijns H, et al. Local adenovirus-mediated transfer of human endothelial nitric oxide synthase reduces luminal narrowing after coronary angioplasty in pigs. Circulation 1998; 98(9): 919–926.
40. Janssens S, Flaherty D, Nong Z, et al. Human endothelial nitric oxide synthase gene transfer inhibits vascular smooth muscle cell proliferation and neointima formation after balloon injury in rats. Circulation 1998; 97: 1274–1281.

Bei Fragen zur Produktsicherheit wenden Sie sich bitte an:
If you have any questions regarding product safety,
please contact:

Walter de Gruyter GmbH
Genthiner Straße 13
10785 Berlin
productsafety@degruyterbrill.com